Lexicon of Terms Relating to the Assessment and Classification of Coal Resources

Lexicon of Terms Relating to the Assessment and Classification of Coal Resources

Arthur H J Todd
Formerly Senior Geologist, World Coal Resources & Reserves Data
Bank Service, IEA Coal Research, London

Graham & Trotman Limited
Publishers

First published in 1982 by

Graham & Trotman Limited
Sterling House
66 Wilton Road
London SW1V 1DE

© IEA Coal Research 1982

14/15 Lower Grosvenor Place
London SW1W 0EX

British Library Cataloguing in Publication Data

Todd, Arthur H. J.
Lexicon of terms relating to the assessment and classification of coal resources.
1. Coal — Dictionaries
I. Title
333.8′22 0321   TN800

ISBN 0-86010-403-6

ISBN 0 86010 403 6

Neither the World Coal Resources and Reserves Data Bank Service nor any supporting country or organisation, nor any employees or contractors of NCB (IEA Services) Ltd, makes any warranty, expressed or implied, or assumes any liability or responsibility for the accuracy, completeness or usefulness of any information, apparatus, product or process disclosed, or represents that its use would not infringe privately-owned rights.

This publication is protected by International Copyright Law. All rights reserved. No part of this publication may be reproduced, stored in a retrieval system or tranmitted in any form, or by any means electronic, mechanical, photocopying, recording or otherwise, without the prior permission of the copyright holder.

Printed in Great Britain by Robert Hartnoll Ltd, Bodmin, Cornwall

Contents

Preface 7
Acknowledgements 9
Lexicon 11
Sources 134

PREFACE

In the early 1970's, the International Energy Agency (IEA) was formed in response to the perceived threat to future energy supplies posed by the "Oil Crisis". In addition to measures designed to combat immediate threats to the flow of oil supplies to member countries it was decided to institute a programme of research aimed towards ensuring the future supply of other fuels in substitution for any missing sources of energy supply. More than fifty projects were set up, the work being shared among the various member countries of the IEA.

This Lexicon on coal resources and reserves has been produced by the World Coal Resources and Reserves Data Bank Service, one of four coal information services operated by IEA Coal Research to assist those involved in many aspects of the "supply chain" for coal, particularly in relation to bulk energy supply. Those who undertook this study of Coal Resources and Reserves very soon realised there are many different conventions in use for the description and assessment of coal deposits, making valid comparisons difficult. It was therefore decided that a study should be made of all available conventions, with a view to better understanding, and possibly translation from one convention to another. As the study progressed it became clear that, even if pursued to the limit, no national convention would be found which provided figures capable of being integrated into the international supply chain for coal; a different methodology would have to be developed. However, since all publicly available data on the world's coal resources and reserves are expressed in terms of the conventions currently in use, it is at least necessary to know what the terms mean in each case.

All of the terms found during this study have been abstracted from published standards and conventions, and where necessary translated into English. These have been arranged in alphabetical order and in each case the original source is indicated by an abbreviated reference, a full list of sources being given in an appendix.

This work was originally carried out for internal use as part of the activity of the Resources and Reserves Project. It is now being made available to a much wider public as it is felt that the information will be of use and value to those involved in coal trade as well as to research workers and energy resources analysts

Only those terms and definitions which are precisely defined in authorative documents have been included in this Lexicon. In recent years there has been active international co-operation towards the establishment of standards for petrographic analysis of bituminous coal and anthracite, and the draft ISO/DIS 7404 has now been circulated for final comment and approval. This document includes a glossary of terms together with methods of sample preparation and the determination of maceral composition microlithotype composition and vitrinite reflectance. Such a document can make a valuable contribution to worldwide uniformity in coal description.

ACKNOWLEDGEMENTS

IEA Coal Research wish to acknowledge the many authorities who have kindly agreed to the reproduction of extracts from their publications. Thanks are due to the following organisations:

American Society for Testing and Materials; Belgium Coal Mining Industry Executive; British Standards Institution; Charbonnages de France Technical Services, Directorate; Department of Mines Queensland; Department of Mines, Republic of South Africa; Energy Mines and Resources Department, Government of Canada; Eni-Enel-Finsider; Federal Ministry for Commerce, Trade and Industry Austria, Geological Survey of Queensland; Geological Survey of Taiwan; German Institute of Standards; Indian Standards Institution; International Organisation for Standardisation; Japanese Standards Association; National Coal Board; New Zealand Geological Survey; Polish Committee for Standardisation; Rhurkohle A. G.; Standards Association of Australia; Standing Committee on Coalfield Geology of New South Wales; United Nations, Committee on Natural Resources; United Nations, Economic Commission for Europe; US Bureau of Mines, US Geological Survey.

The majority of source documents were in English. Where this was not so, translations were made by or on behalf of World Coal Resources and Reserves Data Bank Service. Extracts from DIN Standards were translated by a member of the project team Mr. G. Dennett; these translations have not been checked by DIN Deutsches Institut für Normung e.V., Berlin. The rest of the translations were done on a similar basis by other professional translators.

A

abandoned mines - : See **deposit access status** - Austrian guidelines.

abbaufähige Vorräte - German term: See **mining resources** - FRG standard procedure.

Abbaumächtigkeit - German term: See **worked thickness** - CEC Analysis of terms.

Abbauverluste - : German term: See **mining losses** - FRG standard procedure.

Abbauvorrat - German term: See **working reserves** - CEC Analysis of terms.

abbauwürdig - : German term: See **economic** - Austrian guidelines.

Abbauwürdigkeit - German term: See **economic mineability** - Austrian guidelines.

abnormal coals - Term in Indian standard procedure: Some coals may exhibit abnormal properties because of their unusual petrographic composition e.g. some coal may show low moisture content but at the same time be non-coking. These should preferably be reported separately. (Indian Standard Procedure for Coal Reserve Estimation, 1977).

absolute joule - Indian standard definition: The basic unit, equivalent to 10.10^6 ergs. [IS: 1350 (Part II) - 1970].

access overburden ratio (Zugangsverhältnis) - FRG standard procedure: For brown coal, this is always got from quantity calculations. It indicates how much access to winnable coal in the surface mine has been achieved or is to be expected from an actual or predicted overburden stripping performance. The same guidelines are used in calculating the access ratio as are used in calculating the operating overburden ratio. (DIN 21 942, 1961).

actual-yield coal reserves - Japanese standard definition: These are reserves obtained through conduct of actual mining operations in **safe coal reserves**. The ratio between these reserves and safe coal reserves, expressed as a percentage, is known as **actual-yield factor**. (JIS M 1002 - 1978).

actual-yield factor - : See **actual-yield coal reserves** - Japanese standard definition.

agglomerating - : See also **caking** and **coking**.

agglomerating - US standard definition: "Coals which in the volatile matter determination produce either an agglomerate button that will support a 500 g weight without pulverizing, or a button showing swelling or cell structure, shall be considered agglomerating from the standpoint of classification". (ANSI/ASTM D 388 - 77, para. 7.2).

A

agglutinant - French term: See **caking** - UN ECE definitions.

air-dried - : See **moisture in air-dried coal**.

air dried coal - Indian standard definition: "The sample of coal which for the purposes of analysis, is exposed to the atmosphere of the laboratory to bring it in equilibrium with the humidity conditions prevailing there, so that the sample does not lose or gain weight during weighing". [IS: 1350 (Part II) - 1970].

air-dry, ash-free (lufttrocken und aschefrei) - FRG standard definition: The air-dry, ash-free condition is a theoretical concept which is applied in the International Classification System for hard coals. The weight of ash remaining after combustion at (815 ± 10)°C is subtracted from the weight of fuel which has been stored at 30°C and 97% relative atmospheric humidity until constant weight is reached. (DIN 51 700, 1967, para. 5.2).

air drying - US standard definition: A process of partial drying of coal to bring it to near equilibrium with the atmosphere in the room in which further **reduction/division** of the sample is to take place. (ASTM D 3302 - 74).

air dry loss - US standard definition: The loss in weight, expressed in percentage, resulting from the partial drying of coal at each stage of **reduction or division**. (ASTM D 3302 - 74).

A:K (Abraum:Kohle) - German term: See **operating thickness overburden ratio** - FRG standard definition.

alginite - US standard definition: A maceral derived from waxy walls (thimble cups) of Botryococcus or allied types of algae. (ANSI/ASTM D 2796 - 77).

an - Abbreviation of German term, analysenfeucht, i.e. analysis moist: See **moisture in analysis sample** - FRG standard definition.

an - US standard abbreviation for **anthracite**, "where it is desired to abbreviate the designation of the ranks of coal by group." (ANSI/ASTM D 388 - 77).

analysenfeucht - Term in FRG standard meaning analysis moist: See **moisture in analysis sample** - FRG standard definition.

analysis - UK standard definition: A quantitative statement of the experimentally determined physical and chemical characteristics of a coal. (BS 3323 : 1978).

analysis, air-dried basis - UK standard definition: An analysis expressed on the basis of a coal sample with moisture content in equilibrium with the ambient laboratory atmosphere. (BS 3323 : 1978).

analysis, dry-ash-free basis (DAF) - UK standard definition: An analysis expressed on a basic condition in which the coal is assumed to be free of both total moisture and ash. (BS 3323 : 1978).

analysis, dry-mineral-matter-free basis (DMMF) - UK standard definition: An analysis expressed on a basic condition in which the coal is assumed to be free of both total moisture and mineral matter. (BS 3323 : 1978).

analysis moist, analysis moisture - : See **moisture in analysis sample**.

analysis sample - FRG standard definition: Analysis sample or **fine sample**, serving for analytical investigations after further preparation of the **laboratory sample** and crushing to analysis size (fineness). See **average sample** for context. (DIN 51 701, 1950, para. 2).

analysis sample - ISO definition: The sample crushed to pass a sieve of 0.2 mm, used for general analysis. (ISO/R 1213/II - 1971, 2.18).

analysis sample - US standard definition: Final subsample prepared from the original gross sample but reduced to 100 percent through No. 60 (250 μm) sieve and divided to not less than 50 g. (ANSI/ASTM D 2013 - 1972, Para. 4.2, and ANSI/ASTM 2234 - 76).

ankerite - UK standard definition: A white mineral, sometimes discoloured, which commonly occurs in the partings of coal. The main constituents are calcium, magnesium and iron carbonates. (BS 3323 : 1978).

ankerite fusain - : See **parting** - ISO definition.

anthracite - : See also **low volatile coal**.

anthracite - Indian standard classification:

CLASS and Type.	Symbol	Nature	Gross cv kcal/kg (dmf)	Volatile Matter Percent (dmf)	G - K Coke Type	Moisture (60% RH) Part/100 parts Unit Coat
ANTHRACITE						
Semi-Anthracite	SA	Non-caking	8250 to 8700	10 to 15	A	< 2
Anthracite	A	Non-caking	8500 to 8700	< 10	A	-

Note. Based on available data, broad ranges of the properties are given. (IS: 770 - 1977).

anthracite - NCB, UK, definition:

NCB Coal Rank Code		Volatile Matter (d.m.m.f.) (per cent)	Gray-King * Coke Type
Main Class	Class		
100		Under 9.1	)
	101 **	Under 6.1	) A
	102 **	6.1 - 9.0	)

* Coals with volatile matter of under 19.6% are classified by using the parameter of volatile matter alone; the Gray-King coke types quoted for these coals indicate the general ranges found in practice, and are not criteria for classification.

** In order to divide anthracites into two classes, it is sometimes convenient

A

to use a hydrogen content of 3.35 per cent. (dmmf) instead of a volatile matter of 6.0 per cent. as the limiting criterion. In the original Coal Survey rank coding system the anthracites were divided into four classes then designated 101, 102, 103 and 104. Although the present division into two classes satisfies most requirements it may sometimes be necessary to recognise more than two classes. (NCB Coal Classification System, 1964).

anthracite - UK standard definition: Coal of the highest rank, with a semi-metallic lustre and capable of smoke-free combustion. NOTE: It corresponds to NCB code numbers 101 and 102. (BS 3323 : 1978).

anthracite - US standard classification:

Class	Group	Fixed Carbon Limits, percent (Dry, Mineral-Matter-Free Basis)		Volatile Matter Limits, percent (Dry, mineral-Matter-Free Basis)		Calorific Value Limits Btu per pound (Moist,[1] Mineral-Matter-Free Basis)		Agglomerating Character
		Equal or Greater Than	Less Than	Greater Than	Equal or Less Than	Equal or Greater Than	Less Than	
1.Anthracite	1.Meta-anthracite	98	...	...	2	...	...	non-agglomerating
	2.Anthracite	92	98	2	8	...	...	
	3.Semi-anthracite	86	92	8	14	...	...	

* If agglomerating, classify in low-volatile group of the bituminous class. (ANSI/ASTM D 388 - 77, Table 1).

anthracite (Anthrazit) - RAG, FRG, definition: V.M. (daf) by weight: 7% to 10%; nature of crucible coke: powdery. (Ruhrkohlen-Handbuch, 1969, para. 1.1.1.1, Table).

anthracite (Anthrazit, Meta-Anthrazit) - FRG standard definition:

Name (indication)	International Class (First Digit in index no.)	Volatile Matter Content in dry % ash-free matter by weight
Meta-Anthracite) Meta-Anthrazit)	0	0 to 3
Anthracite)) Anthrazit)	1 A B	3 to 6.5 6.5 to 10

(DIN 23 003, 1976, para. 2.1).

anthracite (antraciet) - Belgian definition: Less than 10% V.M.; anthracite B/antraciet B: from 10% to less than 12% V.M. (Analysis to a basis of dry coal containing 5% ash). ("Moniteur Belge-Belgisch Staatsblad" 19 August 1971).

anthracite (antraciti) - Italian definition: Has a volatile matter content from 2% to 10% on a dry ash-free basis, within the general classification of **hard coal**. (ENI-ENEL-FINSIDER Classification Systems for Coals, 1978).

A

anthracite (antracyt) - Polish standard classification:

Coal Type		Classification criteria			
Name	Code Number	Volatile Matter (daf) %	Caking Properties Roga Index	Dilatation	Heat of combustion (daf) kcal/kg
Anthracite Antracyt	42	3-10	0	Not standardised	Not standardised

General technical characteristics: Very low volatile matter content; no caking properties.
Volatile matter determined according to Polish standard PN-71G-04516 (Hard coal);
Caking properties determined according to Polish standard PN-69/G-04518 (Hard coal). (PN-68/G-97002).

Anthrazit - German term: See **anthracite** - FRG standard and RAG, FRG, definitions.

antraciet - Flemish term: See **anthracite** - Belgian definition.

antraciti - Italian term: See **anthracite** - Italian definition.

antracyt - Polish term: See **anthracite** - Polish standard classification.

antracytowy wegiel - Polish term for anthracitic coal: - See **low volatile coals** - Polish standard classification.

A"r": K"r" (Abraum"r" : Kohle"r") - German term: See **operating overburden ratio** - FRG standard definition.

area limiting criteria - NCB, UK, procedure: From the assessment will be excluded areas of coal in which a seam is:-
(i)   Over 1200 m deep;
(ii)   Too tectonically disturbed to work;
(iii)   Too thin to work (seams under 60 cm thick to be included only when of special quality or customarily worked);
(iv)   Too poor in quality to work (ash and/or sulphur, etc. in coal too high, or ratio of coal to dirt too low);
(v)   Too close to an adjacent seam, which has been included in the assessment, for both to be worked;
(vi)   Too badly affected by underworking or overworking for the mining system to be employed:
(vii)   Too wet to work, by proximity either to a natural aquifer or to adjacent flooded workings:
(viii) Required to support property and surface installations, or to maintain land drainage;
(ix)   Likely to be left unworked because of economic layout of workings.

Reserves zones should be delimited wherever practicable by pre-existing boundaries such as faults, margins of zones of impoverishment of the seam, etc. which form natural mining barriers. (NCB, UK, Procedure for the Assessment of Reserves, 1972).

area limiting criteria - Ohai coalfield, NZ: These are as given for the Reefton

A

coalfield except that a) all seam outcrop areas are excluded regardless of dip, and b) faults with vertical displacements greater than 50 ft (instead of 100 ft) are used as boundaries. (NZ geol. Surv. Bull. No. 51, 1964).

area limiting criteria - Reefton coalfield, NZ: Estimate areas are bounded by:
(1) A change from an underground mining estimate area to an opencast estimate area.
(2) The outcrop of the roof of a seam. Coal in the plan area between the outcrop of roof and floor of the seam is excluded from the estimates, except where the seam dips at 70° or more. Much of this excluded coal will be weathered.
(3) The seam becoming unworkable (see **thickness, depth, dip** and **ash limiting criteria**)
(4) A fault with a throw greater than 100 ft., unless the seam on both sides is easily accessible.
Although faults with less than 100 ft. throw may be sufficient to limit the mining area, particularly where the capital available is small, in many mines such faults have been successfully penetrated. Larger faults are only exceptionnaly penetrated within one set of workings, and should be used as the boundaries between estimate areas: (NZ geol. Surv. Bull. No. 56, 1957).

area measurements - FRG standard procedure: Seam parts which are to have their coal resources calculated are divided into sections of which one can expect there to be uniform development from the probable variation as shown on the map of seam properties. For each one of these sections the seam area is taken from the tectonic map and the expected average "mined thickness" from the structural map. (DIN 21941, 1953).

Asche - German term: See **ash** - FRG standard definition.

as delivered - : See **raw** - FRG standard definition.

as-determined basis - US standard definition: Analytical data obtained from the **analysis sample** of coal or coke after conditioning and preparation to No. 60 (250 μm) sieve in accordance with Method D 2013. As-determined data represent the numerical values obtained at the particular moisture level in the sample at the time of analysis. These values are normally converted to conventional reporting bases. (ASTM D 3180 - 74).

ash - Australian standard definition: The inorganic residue after the incineration of coal to constant weight under standard conditions. (AS K 184 - 1969, para. 3.6).

ash (Asche) - FRG standard definition: The residue after combustion at a temperature of 815°C is known as ash. (DIN 51 719, 1978).

ash - Indian standard definition: "Inorganic residue left when coal or coke is incinerated in air to constant weight under specified conditions". [IS 1350 (Part I) - 1969].

ash - UK standard definition: The inorganic matter remaining after the coal has been incinerated to constant mass under standardized conditions. (BS 1016 : Part 3 : 1973).

ash - US standard definition: Inorganic residue remaining after ignition of combustible substances, determined by definite prescribed methods.
Note 1 - Ash may not be identical, in composition or quantity, with the inorganic substances present in the material before ignition.

Note 2 - In the case of coal and coke, the methods employed shall be those prescribed in ASTM Method D 3174, Test for Ash in the analysis sample of Coal and Coke. (ANSI/ASTM D 121 - 76).

ash limiting criteria - India: Bituminous coal reserves are subclassified by ash content ranging up to a maximum of 50% ash for **low** to **medium volatile coals** or **coking coals** and up to a maximum of 55% of ash-moisture for **high volatile** or high moisture **coals**. See **reserves, subclassifications**. (Indian Standard Procedure for Coal Reserve Estimation, 1977).

ash limiting criteria - NCB, UK, procedure: From the assessment will be excluded areas of coal in which a seam is too poor in quality to work (ash and/or sulphur, etc in coal too high, or ratio of coal to dirt too low) (NCB, Procedure for the Assessment of Reserves, 1972).

ash limiting criteria - Reefton coalfield, NZ: No coal with a known ash content of more than 10 per cent is included in an estimate area. This limit of 10 per cent is arbitrary because it is by no means certain that coal with a lower ash content will be marketable or that coal with a higher ash content will not. Much depends on the form of the ash - whether it is obvious in partings in the seam that cannot be excluded in mining, or less obvious as non-combustible material finely disseminated throughout the coal. (NZ geol. Surv. Bull. No. 56, 1957).

ash limiting criteria - USA: Coal containing more than 33 percent ash is excluded from resource and reserve estimates of the Department of the Interior, from January 1, 1975. (U.S. geol. Surv. Bull. 1450-B, 1976).

as-received basis - US standard definition: Analytical data calculated to the moisture condition of the sample as it arrived at the laboratory and before any processing or conditioning. If the sample has been maintained in a sealed state so that there has been no gain or loss, the as-received basis is equivalent to the moisture basis as sampled. (ASTM D 3180 - 74).

assumed reserves - NSW, Australia, definition:- are those which are assumed from general geological conditions and sparse information, and for which it could be expected that most would be raised to a higher category with additional information.
1. The term is intended for use within colliery holdings or within an area of about the size of a holding or even a group of holdings.
2. Generally the points of observation are spaced not greater than 4 km apart. At this spacing, the assumption of continuity of coal must be based largely on geological inference. (NSW Code for Coal Reserves, 1979).

assurance - Austrian guidelines: Resources are assigned to **resource groups** A, B, C1 and C2 when additional exposures do not cause a falling short of their stated limits of accuracy. The required density of exploration work for classification in the above groups must be determined separately from deposit to deposit in accordance with previous experience and must be substantiated if necessary. (Austrian Guidelines for Coal Deposit Assessment, 1972).

attrital coal - US standard definition: The ground mass or matrix of **banded coal** in which **vitrain** and, commonly, **fusain** bands as well, are embedded or enclosed. Note - Attrital coal and other layers in banded coal which give the banded appearance are commonly 1 to 30 mm thick. Attrital coal in banded coal is highly varied in composition and appearance, its luster varying from a brilliance nearly equal to that of the associated vitrain to nearly as dull as fusain; it exhibits striated, granulose or rough texture. In a few cases, rel-

A

atively thick layers of such attrital coal have been found that contain no interbanded vitrain. **Nonbanded coal** also is attrital coal but it is not usually referred to as such. In contrast to the coarser and more variable texture of attrital coal in banded coal, nonbanded coal is notably uniform and fine in texture, being derived from size-sorted plant debris. (ANSI/ASTM D 2796 - 77).

Audibert-Arnu test - UN (ECE) definition: The object of the test is to determine the **coking** properties of coal or coal blends on the laboratory scale. It is not designed, however, to indicate the pressures exerted by coals on the walls of industrial carbonisation ovens. A pencil made of powdered coal is inserted in a well-calibrated narrow tube and topped by a steel rod (piston) which slides in the bore of the tube. The whole is heated at a constant and definite rate. By making regular readings of the displacement of the piston as a function of the temperature, and expressing the displacements observed as percentages of the original length of the pencil, a curve ... can be plotted. (International Classification of Hard Coal by Type, 1956, app. III).

Ausbeute - German term: See **yield** - CEC Analysis of Terms.

Ausbeutefaktor - German term: See **recovery factor** - CEC Analysis of Terms.

average sample - FRG standard definition: An average sample is a portion of a fuel corresponding in respect to properties and size to the average of the whole quantity to be judged in a proportionate way. It is formed by individual samples the collecting of which must be evenly distributed over the whole quantity of fuel. In the sense of this definition the following average samples are distinguished: **rough sample, laboratory sample** or **coarse sample, analysis sample** or **fine sample**, and **moisture sample**. (DIN 51 701, 1950, para. 2).

average thickness (mittlere Mächtigkeit; puissance moyenne) - CEC Analysis of Terms: Simple arithmetic mean of thicknesses approximately homogeneously distributed over an area. (CEC Assessment of Coal Reserves, 1980).

B

backend - German term: See **caking** - FRG standard definition.

Ballast - German term: See **inerts** - FRG standard definition.

banded coal - US standard definition: Coal that is visibly heterogeneous in composition, being composed of layers of **vitrain** and **attrital** coal and, commonly, **fusain**. (ANSI/ASTM D 2796 - 77).

banded coal - UK standard definition: Coal composed of dull and bright layers. (BS 3323 : 1978).

bauwürdig - German term: See **economic** - FRG standard definition.

bedingt abbauwürdig - German term: See **subeconomic** - Austrian guidelines and FRG definition.

bed moisture - : See **inherent moisture (in coal)** - US standard definition.

bergbauliche Vorräte - German term: See **mineable resources** - FRG standard definition.

Bergemittel, Mittel - German terms: See **dirt band** - CEC Analysis of Terms.

bituminous coal - UK standard definition: A general descriptive term for coal between **low volatile coal** and **lignite**. Note: It corresponds to NCB code numbers 800 to 900 inclusive. (BS 3323 : 1978).

bituminous coal - US standard definition: See relevant entries against **high, low** and **medium volatile coals**.

bituminous coal (litantraci) - Italian definitions:

Coal Class Name	Content of V.M. (d.a.f.) %	Gross Cal. Val. moist, without ash kcal/kg.	Index of reflectivity of vitrinite %
Litantraci basso volatili Litantraci medio volatili Litantraci alto volatili A Litantraci alto volatili B	10 - 20 20 - 28 28 - 42 > 42	> 5700	> 1.5 1.1 to 1.5 > 1.1

(ENI-ENEL-FINSIDER Classification Systems for Coal, 1978).

B

bituminous coal - Indian standard classification:

Type	Subdivision or group Name	Group Symbol	Range of Volatiles Percent at 900° 15°C (Unit Coal Basis)	Range of Gross Calorific Value Kcal/Kg (Unit Coal Basis)	Range of Moisture Percent (Mineral-Free Coal Basis) Near-Saturation at 95 per cent Rh at 40°C	Air-Dried at 60 percent Rh at 40°C	Chief Uses
(1)	(2)	(3)	(4)	(5)	(6)	(7)	(8)
Anthracites	Anthracite	A1	3 to 10	8330 to 8670	2 to 4	1 to 3	Gasification process
Bituminous coals (caking) strength increasing from B3 to B1	Low volatile (caking)	B1	15 to 20	8670 to 8890	1.5 to 2.5	0.5 to 1.5	Carbonization for metallurgical coke. Typical coking coals.
	medium volatile (caking)	B1	20 to 32	8440 to 8780	1.5 to 2.5	0.5 to 2	
	High volatile (caking)	B1	Over 32	8280 to 8610	2 to 5	1 to 3	Coking coals, gas coals, gasification.
	High volatile (semi-caking)	B1	Over 32	8060 to 8440	5 to 10	3 to 7	Gas coals, gasification, long flame.
	High volatile (non-caking)	B1	Over 32	7500 to 8060	10 to 20	7 to 14	Steam-raising gasification. Long flame heating.
Sub-bituminous coals	Non-caking, slacking on weathering	B1	Over 32	6940 to 7500	20 to 30	10 to 20	Steam raising and gasification.
Lignites or brown coals	Normal lignite	L1	45 to 55	6110 to 6910	30 to 70	10 to 25	Steam raising, briquetting, gasification distilation.
		L1	55 to 65	6940 to 7500	30 to 70	10 to 25	

(IS: 770 - 1977).

Blähzahl - German term for **swelling index** : See **crucible swelling number** - FRG standard definition.

boghead coal - UK standard definition: Coal resembling **cannel coal** in physical appearance and properties, but distinguished microscopically by the presence of the remains of algae. (BS 3323 : 1978).

boghead coal - US standard definition: **Nonbanded coal in which the exinite (the waxy component) is predominantly alginite.** Note: Transitions between **cannel** and boghead, that is, coals containing both types of exinite, are also known. Microscopical examination is essential for differentiation of the two kinds of non-banded coal and their transitions. (ANSI/ASTM D 2796 - 77).

bone coal - US standard definition: **Impure coal that contains clay or other fine-grained detrital mineral matter.** (ANSI/ASTM D 2796 - 77).

Brennwert - German term: See **gross calorific value** - FRG standard definition.

Brennwert (spezifischer) - German term: See **gross (specific) calorific value** - CEC Analysis of Terms.

bright coal - UK standard definition: Coal of a bright, lustrous appearance, consisting predominantly of **vitrain** and **clarain**. (BS 3323 : 1978).

brown coal : See **lignite**, with which this term is often used interchangeably.

brown coal - UK standard definition: Coal of the lowest rank, which is of a soft friable nature and has a high inherent moisture content. (BS 3323 : 1978).

Bruttomächtigkeit - German term: See **gross thickness** - CEC Analysis of Terms.

burnt out coal - Indian standard procedure: The burnt out portions of coal and jhama shall be excluded while taking thickness of the seams for the purpose of calculation. (Indian Standard Procedure for Coal Reserve Estimation, 1977).

C

caking (backend) - FRG standard definition: A **hard coal** without caking capacity does not soften (under heating) and leaves a powdery coking residue. A hard coal with caking capacity softens and is more or less inflated by gases released by the heat. A destructive distillation (in a crucible) progresses the coal which has become plastic cakes to a crucible coke of a distinct shape... (DIN 51 741, para. 3).

caking (agglutinant) - UN ECE definitions: i) caking properties... in broad terms reflect the behaviour of a coal when it is heated rapidly, as for example in combustion processes. ii) By the caking power of coal we understand that property which results in the formation of fused coke when comminuted coal is degasified at high temperatures. (International Classification of Hard Coals Type, 1956, ch. II.2).

caking coal - UK standard definition: Coal that leaves a coherent residue when carbonized. (BS 3323 : 1978).

calorific value - : See **gross calorific value; net calorific value**.

cannel coal - UK standard definition: Strong, **nonbanded coal** of satin sheen or waxy lustre, showing a conchoidal fracture. It is generally high in **volatile matter** and readily ignited. (BS 3323 : 1978).

cannel coal - US standard definition: **Nonbanded coal** in which the **exinite** is predominantly **sporinite**. Note: Transitions between cannel and **boghead**, that is, coals containing both types of exinite, are also known. Microscopical examination is essential for differentiation of the two kinds of nonbanded coal and their transitions. (ANSI/ASTM D 2796 - 77).

carboni - Italian term: See **hard coal** - Italian definition.

carboni fossili - Italian term: See **coals** - Italian definition.

catégories de charbon - French term: See **classes of coal** - CEC Analysis of Terms.

category "a" reserves - CdF, France, classification: Relating to degree of certtainty, these are **proved reserves** where exploration is so advanced that there is no doubt as to their existence. In a favourable situation where the measures are of regular composition and **dip**, and contain few faults, exploration may simply amount to guarantees as to the continuity and composition of the measures, obtained by drilling or cutting through seams by means of cross cuts. These guarantees will only be valid a short distance from the seam explorations and cuttings and if the density of the drillings or seam cuttings is judged to be sufficiently compact. (CdF Reserves Classification, 1972 Model).

C

category "b" reserves - CdF, France, classification: Relating to degree of certainty:
1. In the levels being worked and the levels immediately below, these consist of reserves not yet explored but whose existence is assessed favourably. Near worked zones (distance from such zones varies according to the regularity of the measures), continuation of measures is ascertained by cross cutting or drilling etc.
2. In deep levels, these consist of resources whose existence and regularity of structure have been sufficiently explored by drilling so that the presence assessment is favourable. (CdF Reserves Classification, 1972 Model).

category "c" reserves - CdF, France, classification: Relating to degree of certtainty, these consist of reserves in zones to which the measures and seams could logically extend but where there is only inadequate proof if any. (CdF Reserves Classification, 1972 Model).

category R-1 - UN CNR definition: This category encompasses the in situ resources in deposits that have been examined in sufficient detail to establish their mode of occurrence, size and essential qualities within individual deposits and the distribution of quality and the physical properties that affect mining, are known mainly by direct physical penetration and measurement of the deposit combined with limited extrapolation of geological, geophysical and geochemical data. Quantities have been estimated at a relatively high level of assurance, although in some deposits the estimation error may be as high as 50 per cent. The primary relevance of such estimates is in the planning of mining activities. This category can be further subdivided into **subcategories E, S, and M**. Some of the more common terms that, to varying degree, have been used for this category are: established, **demonstrated** reasonably, assured. explored. (UN CNR International Classification of Mineral Resources, 1979).

category r-1 - UN CNR definition: This designates the corresponding **recoverable resource** for category R-1 of in situ **resources** and may correspondingly be subdivided into **subcategories E, S, and M**, where warranted. (UN CNR International Classification of Mineral Resources, 1979).

category R-2 - UN CNR definition: This category provides for estimates of in situ **resources** that are directly associated with discovered mineral deposits but, unlike the resources included in category R-1, the estimates are preliminary and based largely upon broad geological knowledge supported by measurements at some points. The mode of occurrence, size and shape are inferred by analogy with nearby deposits included in R-1, by general geological and structural considerations and by analysis of direct or indirect indications of mineral deposition. Less reliance can be placed on estimates of quantities in this category than on those in R-1; estimation errors may be greater than 50 per cent. The estimates in R-2 are relevant mostly for planning further exploration with an expectation of eventual reclassification to category R-1. If warranted. this category may be further subdivided into **subcategories** E and S. Some of the more common terms that, to varying degree, have been used for this category are: **inferred**, estimated additional, **possible**. (UN CNR International Classification of Mineral Resources, 1979).

category r-2 - UN CNR definition: This designates the corresponding **recoverable resource** for category R-2 of in situ **resources** and may correspondingly be subdivided into **subcategories E, S, and M**, where warranted. (UN CNR International Classification of Mineral Resources, 1979).

category R-3 - UN CNR definition: These **resources** are undisovered but are thought to exist in discoverable deposits of generally recognised types. Esti-

C

mates of in situ quantities are made mostly on the basis of geological extrapolation, geophysical or geochemical indications, or statistical analogy. The existence and size of any deposits in this category are necessarily speculative. They may or may not actually be discovered within the next few decades. Estimates for R-3 suggest the extent of exploration opportunities and the somewhat longer range prospects for raw material supply. Their low degree of reliability should be reflected by reporting in ranges. Some of the more common terms that, to varying degree, have been used for this category are: undiscovered, potential, hypothetical, speculative, prognostic. (UN CNR International Classification of Mineral Resources, 1979).

category r-3 - UN CNR definition: This designates the corresponding **recoverable resource** for **category R-3** of in situ **resources** and may correspondingly be subdivided into **subcategories E, S, and M**, where warranted. (UN CNR International Classification of Mineral Resources, 1979).

certain (zekere/certaines) reserves - NVKS, Belgium, classification: Applicable in the leased areas of the Kempen/Campine coalfield to **recoverable** coal from seams known well enough to determine the the workability of the seam in terms of the thickness of coal and the dirt bands within the seam, as well as the structural trend and the conditions of the roof and the floor, they must be known from previous workings in the same seam and from nearby underground workings or from extensive development or exploratory drivages to qualify for this degree of assurance. They represent 90% of the corresponding coal in place in layout plans. (Belg. Coal Min. Ind. Exec., 1963, and NVKS, 1978).

certaines reserves - : See **certain reserves** - NVKS, Belgium classification.

charbon - French term: See **coal (hard coal)** - CEC Analysis of Terms.

chudy wegiel - Polish term for lean coal: - See **medium volatile coal** - Polish standard classification.

cinder coal - UK standard definition: Coal that has undergone a severe form of heat alteration in which the coal has been partially or totally carbonized in situ. (BS 3323 : 1978).

clarain - UK standard definition: A bright coal, lustrous but not glassy in appearance, which may be very finely striated. (BS 3323 : 1978).

Class I reserves - NCB, UK, definition: These are workable reserves for which the density of geological information is sufficient to give accurate information on variations of seam thickness, depths, gradients, quality, and other relevant geological factors and so enable a reliable estimate to be prepared of the total tonnage capable of supporting extraction in conformity with the Action Programme and the long-term development plan for the colliery. This implies that:-
(i) the degree of geological uncertainty has been so reduced that the only geological hazards that will be encountered are those whose location and effect are sufficiently predictable to avoid any significant variations in the quality of the product or material shortfall in the planned production from the reserves concerned, taking into account the effective spare productive capacity of the colliery;
(ii) the rate of production which is expected of the mining method to be used has already been established in a comparable geological environment.
Geological conditions determining classification:
(i) Sedimentary environment.
In general, for an area of reserves to be allocated to Class I, knowledge

of the sedimentary environment along at least two sides of this area if it is triangular in shape, or along at least three sides if it is quadrilateral in shape, is necessary from previous workings and/or proving drivages Alternatively, information must be available from a sufficient number of suitably located boreholes to be sure that no unexpected change can arise which would cause significant variation in the quality of the product or material shortfall in the planned production. If there is positive evidence of a low rate of regional variation of all sedimentary phenomena, reserves could be allocated to Class I on a sligtly less full extent of proving provided the full requirements of Class I can still be met.
(ii) Tectonic environment.
An area of reserves should not be designated Class I when the seam has not been overworked or underworked, unless the regional intensity of faulting is very low and there is sufficient knowledge of the effect and location of faults to ensure that no unexpected faulting could cause material shortfall in the planned production. The rate of change of strike must be sufficiently low for valid lateral extrapolation between provings within or around the area.

Mining and economic considerations determining classification:
The Action Programme and long-term development plan for any colliery are based upon the presumed availability and the deployment of specified mining resources (manpower, machinery, materials, etc.) and upon certain assumptions concerning the marketability of and proceeds from the products. It follows that an area of reserves cannot be allocated to Class I, however much is known of its geological conditions, unless the planned level and rate of production is compatible with the conditions, and also unless the mining and economic factors in the Action Programme are comparable with those in an Action Programme whose objective has already been achieved under similar geological conditions. (NCB, UK, Procedure for the Assessment of Reserves, 1972).

Class II reserves - NCB, UK, definition: These are workable reserves for which the density of geological information is sufficient in regard to thickness, depth, gradient, quality and geological environment to support the preparation of a detailed extraction plan, but for which the general level of geological uncertainty regarding the number, location and effect of local hazards is too great to ensure that the forecast production will be achieved and/or that the desired quality of the products will be maintained. This implies that:-
(i) the nature of all the significant geological hazards is known but their extent and location have not been sufficiently determined to be assured that the planned level of production can be maintained, taking into account the effective spare productive capacity of the colliery;
(ii) the rate of production which is expected of the mining method to be used may not already have been established in a comparable geological environment;
(iii) when this class of reserves is being worked the geological situation will require constant review as geological data become available from the production units. The colliery Action Programme may need regular revision to avoid unplanned production shortfalls;
(iv) a high proportion of the reserves in this class should, with further proving, achieve Class I status.

Geological conditions determining classification:
(i) Sedimentary environment.
For an area of reserves to be designated Class II, the sedimentary environment will normally have been proved along at least one side, or sufficient borehole information will be available to guarantee that the nature of all sedimentary phenomena has been determined and that none are of the type or intensity which could give rise to major shortfalls

C

          in planned production.
- (ii) Tectonic environment.
  In Class II the density of geological information on faulting is insufficient to be certain of their location, intensity, or effect, so that some unexpected shortfalls in planned production could occur even with the spare productive capacity provided. (NCB, UK, Procedure for the Assessment of Reserves, 1972).

Class III reserves - NCB, UK definition: These are reserves of which the degree of geological knowledge is such that unpredictable facestopping geological hazards may be encountered and that the requirements regarding the validity of any detailed extraction plan, the maintenance of required production levels and product quality standards governing the higher classes, cannot be met. This implies that:
- (i) the reserves are associated with a high level of geological uncertainty and that the information is insufficient to support the preparation of a reliably phased and detailed layout or of reliable production estimates, so that (unlike Class II) major shortfalls in the level of production could occur;
- (ii) a highly variable proportion of reserves in this class should, with further proving and/or better definition of the future objectives and resources of the mine, achieve higher status;
- (iii) the colliery Action Programme will be unreliable to the extent that this class of reserves is being worked and may require frequent adjustment. The existence of unpredictable hazards will give rise to the need to carry an exceptional amount of spare productive capacity.

Geological conditions determining classification:
- (i) Sedimentary environment.
  In Class III the degree of geological knowledge is insufficient to eliminate the possibility of such adverse sedimentary phenomena as washouts, seam splits, and the margins of areas of regional impoverishment whose occurrence could cause major shortfalls in production and substantial losses in recoverable reserves.
- (ii) Tectonic environment.
  In Class III the degree of geological knowledge on faulting and other tectonic phenomena is such that faces could be stopped prematurely and major shortfalls in production could take place.
- (iii) General: An area where the information points are over two miles apart should normally be excluded from the assessment. (NCB, UK, Procedure for the Assessment of Reserves, 1972).

classes of coal (Steinkohlenklassen; catégories de charbon) - CEC Analysis of Terms: The International Classification of Hard Coals by Type drawn up for the Coal Commission for Europe (ECE) by the Coal Classification Working Party is to be used. Each type of coal is assigned a three-digit code number. The first digit (class) indicates the **volatile matter** content and from the seventh class onwards the **gross calorific value**. The second digit (group) indicates the **swelling index** or caking index (i.e. caking properties) and the third digit (sub-group) indicates **dilatation** behaviour or the Gray-King coke type (i.e. **coking** properties). (CEC Assessment of Coal Reserves, 1980).

clean coal (reserve level 3) - EMR, Canada, definition: Bituminous coal destined for coking coal markets may be upgraded in a cleaning plant to contract specifications, resulting in a final-product tonnage of clean coal. As "free-on-rail" it may represent 75 per cent. to 50 per cent. of recoverable coal. Costs should be reported along with tonnages (or thermal equivalent). (Canada: EMR Report EP 77 - 5, 1977).

clean(ed) coal - ISO definition: Coal which has been treated by a wet or dry cleaning process. (ISO/R 1213/II - 1971, 1.09).

cleaning and preparation losses (perdite di trattamento) - Italian definition: "These are the quantities of **recoverable reserves** which are lost during eventual treatment of mined coal, treatment which varies according to the type of end use of the coal". (ENI-ENEL-FINSIDER Classification Systems for Coals, 1978).

cleat - US standard definition: The joint system of coal beds, usually oriented normal or nearly normal to the bedding. (ANSI/ASTM D 2796 - 77).

coal - NCB, UK, definition: Coal is defined as material with ash not exceeding 15.0%. (NCB, UK, Procedure for the Assessment of Reserves, 1972).

coal - Rep. South Africa, DM, definition: a carbonaceous rock of sedimentary origin containing not more than 50% of ash. (Dept. Mines, Rept. Coal Res. SA 1975).

coal - UK standard definition: A combustible sedimentary rock, formed from altered plant remains consolidated under superimposed strata. The characteristics of different coals are due to differences in source plant material, to the conditions and the degree of change that this has undergone in its geological history, and to the range of impurities present. (BS 3323 : 1978).

coal - US standard definition: A brown to black combustible sedimentary rock (in the geological sense) composed principally of consolidated and chemically altered plant remains. (ANSI/ASTM D 2796 - 77).

coal (hard coal) (Steinkohle; charbon) - CEC Analysis of Terms: Coal with a **gross calorific value** of at least 23900 kJ/kg (air-dried and ash-free). (Corresponding to about 5700 kcal/kg, since 1 kJ = 0.239 kcal. (ISO/R 1928-1971 is to be used to determine the **gross calorific value** and also to convert the gross calorific value into the **net calorific value**. (CEC Assessment of Coal Reserves, 1980).

coal in place (geological reserves) (geologischer Vorrat; ressource géologique - CEC Analysis of Terms: All the carbonaceous matter in the Earth's crust to a specified minimum seam thickness and to a specified maximum depth (generally 0.30 m and 2000 m). Only in a few exceptional cases has this definition of geological resources been of significance hitherto for mining industry. It has therefore hardly ever been determined qualitatively and quantitatively as a separate entity and has been estimated only in regional analyses. (CEC Assessment of Coal Reserves, 1980).

coal in situ - Rep. South Africa, DM, definition: The total amount of coal in a given area occurring in its natural environment; since it includes all coal, however deep or thin it may be, this is an academic figure (see also **degree of certainty, mineable coal in situ**). (Dept. Mines, Rept. Coal Res. SA 1975).

coal preparation - UK standard definition: The physical and mechanical process applied to coals to make them suitable for particular uses. (BS 3323 : 1978).

coal rank code - NCB, UK, classification: The different classes of coal are designated by code numbers termed coal rank codes. The purpose of this system is to describe the **rank**, or degree of **coalification**, of the coal substance of any coal, on the basis of **V.M.** (dmmf) and **Gray-King coke type**. Other properties of the coal as used, or as sold, such as its ash or sulphur content, need to

C

be specified additionally. (NCB Coal Classification System, 1964).

coal resources (risorse de carbon fossile) - Italian definition: These are defined as all those coal deposits in a given area or country which are known or may be tentatively assumed to exist. (ENI-ENEL-FINSIDER Classification Systems for Coals, 1978).

coal substance - UK standard definition: Coal, excluding its mineral matter and moisture. (BS 3323 : 1978).

coalification - UK standard definition: The process by which original plant remains were transformed into coal. This process is characterized by the increasing proportion of the carbon content and the decreasing proportion of volatile matter in the substance on the dry-mineral-matter-free basis. (BS 3323 : 1978).

coals (carboni fossili) - Italian definition:- are those solid fuels produced in the process of coalification of vegetable matter. These are divided into: **hard coal (carboni)**, having a gross calorific value greater than 5700 kcal/kg on the basis of mineral-matterfree coal in equilibrium with air at 30°C and 96% relative humidity; **lignite (ligniti)**, having this parameter less than 5700 kcal/kg. (ENI-ENEL-FINSIDER Classification Systems for Coals, 1978).

coarse sample - : See **laboratory sample** - FRG standard definition.

coefficient calorifique (spécifique) - French term: See **gross (specific) calorific value** - CEC Analysis of Terms.

code number - UK standard definition: A numerical classification index defining a grouping of coals according to **rank** and **caking** properties. (BS 3323 : 1978).

cokefiant - French term: See **coking,** - UN ECE definition.

coking coal - UK standard definition: A coal suitable for carbonization in coke ovens (BS 3323 : 1978).

coking coals - : See also **fat coals, medium volatile coals, high volatile coals**.

coking (cokefiant) - UN ECE definition: Coking properties... in broad terms reflect the behaviour of a coal when it is heated slowly - as, for example, when it is carbonized. (International Classification of Hard Coals by Type, 1956, ch. II.3).

coking steam coals - : See also **low volatile coals**.

concessions - : See **deposit access status,** - Austrian guidelines.

contraction (kontraktion) - in Audibert-Arnu test, FRG standard definition: The percentage decrease in length shown by a coal pencil when it is heated in a tube. (DIN 51 739, 1976, para. 3).

course of dilatation (Dilationsverlauf) - in Audibert-Arnu test, FRG standard definition: The changes in length of a coal pencil with progressive heating under fixed conditions. (DIN 51 739, 1976, para. 3).

crucible swelling number - UK standard definition: The number that, by reference to a series of standard profiles, defines the size and shape of the residue produced when a standard mass of coal is heated under standard conditions. (BS

3323 : 1978).

crucible swelling number or swelling index - Australian standard definition: A number which defines, by reference to a series of standard profiles, the size and shape of the residue produced when a standard weight of coal is heated under standard conditions. (AS K 184 - 1969, para. 3.2).

crucible swelling number or swelling index - UN ECE definition: The coke button obtained by heating the finely ground coal to 820°C - 5°C is classified by comparison with the outlines of a set of standard profiles. The number of the profile most closely corresponding to the coke button obtained is the crucible swelling number. (The International Classification of Hard Coals by Type, 1956, app. I).

crucible swelling number or swelling index (Blähzahl) - FRG standard definition: The swelling index according to the standard is a measurement of the caking capacity of hard coal. The swelling index is established by comparing the coking residue of the sample produced by the procedure laid down in this standard with the swelling index types illustrated. (DIN 51 741, 1974, para. 2).

cumulative depletion - US official definition: "The sum of all coal extracted and lost in mining prior to the date of the estimate. May be subdivided by rank and sub-rank (class and group) of coal, overburden class, thickness class, mining method, heat value, usage, time, cokeability, chemical constituents and area of production". (USGS 1979 Proposed Revision of Bull. 1450-B, 1976).

cumulative production - US official definition: Includes the sum of all production prior to the date of the estimate. (US geol. Surv. Bull. 1450-B, 1976).

cumulative production - US official definition: "The sum of all production from a mine, field, region, state, or nation prior to the date of the estimate. May be subdivided into production by rank and sub-rank (class and group) of coal, overburden class, mining method, heat value, usage, time, cokeability, chemical constituents, and area of production". (USGS 1979 Proposed Revision of Bull. 1450-B, 1976).

cutinite - US standard definition: A maceral derived from the waxy coatings (cuticles) of leaves and other plant parts. (ANSI/ASTM D 2796 - 77).

C.V. - Abbreviation for calorific value: See gross calorific value; net calorific value.

D

DAF, daf - Abbreviation: dry, ash-free.

deducidas, reservas - Spanish term: See **inferred (deduced, or possible) reserves** - Colombian official classification.

deep mineable resources (Tiefbauvorräte) - FRG standard definition: For brown coal, all exploitable resources which are not surface mineable are designated deep mineable resources. (DIN 21 942, 1961).

degré de connaissance - French term: See **degree of exploration** - CEC Analysis of Terms.

degré de sollicitation tectonique - French term: See **degree of tectonic stress** - CEC Analysis of Terms.

degré d'exploration - French term: See **degree of prospecting** - CEC Analysis of Terms.

degré d'investigation - French term: See **degree of investigation** - CEC Analysis of Terms.

degree of certainty (in situ reserves and resources) - Rep. South Africa, term: specifying the minimum number of boreholes and adits per 2,000 hectares for each coalfield for the categories proved, indicated and inferred. Table 5.31 of ref. specifies ranges for each category by coal seam and/or geographical area. (Dept. Mines, Rept. Coal Res. SA 1975).

degree of development - RAG, FRG, guidelines:
i) Developed Coal Reserves: Coal reserves are said to be developed if it is possible to prepare the gate roads.
ii) Undeveloped Coal Reserves: All other coal reserves.
(Ruhrkohle A.G. Guidelines, 1970).

degree of exploration - Austrian guidelines: According to the degree of exploration of the resources they are subdivided and classified in **resource groups A, B, C1 and C2**. (Austrian Guidelines for Coal Deposit Assessment, 1972).

degree of exploration (Untersuchungsgrad; degré de connaissance) - CEC Analysis of Terms: The degree of exploration of a deposit is characterised by the **degree of prospecting** and the **degree of investigation**. (CEC Assessment of Coal Reserves, 1980).

degree of investigation (Erforschungsgrad; degré d'investigation) - CEC Analysis of Terms: Type, extent, quality and documentation of the evaluation of the prospecting work. Evaluation consists of:
- geological evaluation (genesis, stratigraphy, tectonics, petrography, geolo-

D

    gical-engineering and hydrological investigations, etc.),
- raw materials investigations, particularly relating to the deposit (physical, chemical, petrographical).
(CEC Assessment of Coal Reserves, 1980).

degree of prospecting (Erkundungsgrad; degré d'exploration) - CEC Analysis of Terms: Type, extent, spatial distribution, quality and documentation of the prospecting work carried out.
Prospecting work includes:
- stratigraphic surveying,
- geological mapping,
- geophysical measurements,
- sampling,
 in
- surface explorations,
- boreholes,
- underground workings.
(CEC Assessment of Coal Reserves, 1980).

degree of tectonic stress (Grad der tektonischen Beanspruchung; degré de sollicitation tectonique) - CEC Analyis of Terms: Movements in the Earth's crust during or after sedimentation led to new tectonic structures. In deposits these new structures may influence working (micro-tectonics), or hinder working (small- and medium-scale tectonics). The disturbance of coal seams in deposits as a result of tectonic movements therefore has economic consequences. It is therefore significant to describe the degrees of disturbance in terms of tectonic stress. Many attempts have been made to do so, but not all of them have been successful and it has not so far been possible to apply the findings to all deposits. (CEC Assessment of Coal Reserves, 1980).

degrees of exploration - RAG, FRG, guidelines:
i) First Degree: All coal reserves where the tectonics and seam structure are known as a result of exploration. The tectonics are considered to be known if the exploration points are no more than 400 m apart horizontally and/or 200 m apart vertically. The seam structure is considered to be known in an area which is not more than 1000 m away from the exploration point along the strike and along the dip (distances in the plane of the seam).
ii) Second Degree: All coal reserves bordering on the first degree area. The tectonics are considered to be known in this case if the exploration points are no more than 800 m apart horizontally and/or 400 m apart vertically. The seam structure is considered to be known in this case if the distance from the first degree area is no more than 1000 m in the seam. If residual areas are left after drawing the boundaries between i) and ii) they may be allotted to the larger neighbouring area.
iii) Third Degree: All coal reserves bordering on the second degree area which can be determined seam by seam in large tectonic blocks.
iv) Fourth Degree: All other coal reserves in areas where it is impossible to determine reserves accurately seam by seam. (Ruhrkohle A.G. Guidelines, 1970).

demonstrated - US official definition, applicable to both the **reserve** and **identified subeconomic resource** components of the classification system: A collective term for the sum of coal in both **measured** and **indicated resources** and **reserves** (US geol. Surv. Bull. 1450-B, 1976).

demonstrated reserve - US official definition: A collective term designating the sum of **measured** and **indicated reserves**. A tonnage estimate for this category of coal is the sum of the estimates for **measured** and **indicated reserves**.

D

(USGS 1979 Proposed Revision of Bull. 1450-B, 1976).

demonstrated reserve base - US official definition: A tonnage estimate for this category of coal consists of the sum of the estimates for **measured** and **indicated reserves**, marginal reserves and part of the **measured and indicated subeconomic resources** (the coal estimated to be lost in mining). The **demonstrated reserve base** is the same as the **reserve base** which is the preferred usage. (USGS 1979 Proposed Revision of Bull. 1450-B, 1976).

demonstrated reserves (reservas demostradas) - Colombian official classification: This is a classification term which is applied to the sum of **measured** and **indicated reserves**. Many undertakings and finance establishments accept the demonstrated reserve figure for the evaluation of medium and long term projects. (Publ. Geol. Esp. Ingeominas, No. 3, 1979).

demonstrated resource - US official definition: A collective term designating the cumulative sum of the accessed and virgin coal classified as **measured and indicated resources**. These resources are divisible on the basis of economics and criteria of depth, thickness, **rank**, and distance from points of measurement and sampling into **reserves, marginal reserves**, and **subeconomic resources**. A tonnage estimate for this category is the sum of the estimates for the **reserve base and subeconomic resources**". (USGS 1979 Proposed Revision of Bull. 1450-B, 1976).

demostradas, reservas - Spanish term: See **demonstrated reserves** - Colombian official classification.

density fraction (Wichtestufe; tranche densimétrique) - CEC Analysis of Terms: A freely fixable upper limit to density (specific gravity). With mixtures, a density fraction is always fixed if an entire sample is not to be analysed during quality control but only the part whose density is below the density fraction fixed. This part can be eliminated from the sample using physical processes. (CEC Assessment of Coal Reserves, 1980).

deposit access status - Austrian guidelines: Resource distribution by access status: The resources shall be established and presented separately for:
1. Mines in production, covering those deposit resources which are within reach of a mine.
2. Associated reserve occurrences: a) already granted b) not yet granted.
3. Abandoned mines.
4. Other occurrences.

(Austrian Guidelines for Coal Deposit Assessment, 1972).

deposit access status - : See **proved reserves** - Indian standard procedure.

depth (Teufe; profondeur) - CEC Analysis of Terms: A measure of the vertical distance from the surface. Absolute depth is a measure of the vertical distance from a standard level (ordnance datum: in Germany NN = mean sea level). (CEC Assessment of Coal Reserves, 1980).

depth limiting criteria - Belgium: In the State Mining Concession of the Kempen/Campine coalfield of Belgium, the **certain, probable,** and **possible** categories of reserves are assessed to a depth of 1100 m (1000 m below sea level) in the areas leased for mining. Otherwise, **technically workable reserves** are assessed to a depth of 1500 m over the whole area of the State Mining Concession. (Belg. Coal Min. Ind. Exec., 1963, and NVKS, 1978).

depth limiting criteria - Canada:

a) For **resources of immediate interest**, the depth limits vary regionally for various ranks of coal as follows:
i) Cordillera: All coal ranks to depths accessible to surface mining; **anthracite**, and **bituminous coals** to a depth of 1000 ft.
ii) Plains: All coal ranks to depths accessible to surface mining, generally less than 150 ft; **bituminous** and **subbituminous** to depths of 750 ft.
iii) Maritimes: Onshore, to depths of 1000 ft; offshore to depths with not more than 4000 ft of vertical cover.
b) For **resources of future interest**:
i) Cordillera: All coal ranks to depths of 2500 ft;
ii) Plains: All coal ranks to depths of 1500 ft;
iii) Maritimes: All coal ranks onshore to depths of 4500 ft; and within 5 miles offshore, to depths with more than 4000 ft of vertical cover and, beyond 5 miles offshore, to any depth.
(Canada: EMR Report EP 77 - 5, 1977).

depth limiting criteria - CdF, France: With the exception of the Merlebach mine (Lorraine), where reserves are calculated to -1250 m, the maximum depth for calculation purposes is 1200 m. (CdF Reserves Classification, 1972 Model).

depth limiting criteria - Colombia: The depth criteria considered economic by INGEOMINAS, by resource class and degree of metamorphism, are:

Resource Class	Degree of Metamorphism	Depth in m
Total, including potential resources	Anthracite and bituminous coal Subbituminous coal and lignite	1800 or less 1800 or less
Identified resources	Anthracite and bituminous coal Subbituminous coal	1800 or less 1800 or less
Reserve base (geological or "in situ"). Also recoverable (or exploitable) reserves (including the non-recoverable part of the reserve base).	Anthracite and bituminous coal Subbituminous coal Lignite	300 or less 300 or less 40 or less
Subeconomic (potential) resources	Anthracite Bituminous coal Subbituminous coal Lignite	0 to 300 300 to 1800 300 to 1800 40 to 1800

(Publ. Geol. Esp. Ingeominas, No. 3, 1979).

depth limiting criteria - India: Reserve data shall be reported according to depth from the surface down to 1200 m (4000 ft). (Indian Standard Procedure for Coal Reserve Estimation, 1977).

depth limiting criteria - Japanese standard procedure: The present mining depth limit for coal reserves of **verified coal reserves type 1 (A)**, **verified coal reserves type 1 (B)**, **estimated coal reserves type 1**, and **predicted coal reserves**

D

type 1 may not exceed 600 m below mine entrance level for a standard coal seam. Where, however, mine depth partial development exceeding 300 m is considered possible in developed areas, this depth limit may be increased 300 m. The same type of operation is performed for each 300 m mine extension. The mine entrance level specified herein denotes the mine entrance level in developed areas and the predicted mine entrance level in the undeveloped areas.

The future mining depth limit for verified coal reserves type 2, estimated coal reserves type 2, and predicted coal reserves type 2 may not exceed 1200 m below mine entrance level for a standard coal seam.

In the case of coal seams of class 2 and class 3 of the coal thickness classes, individual mining depth limits of standard coal seams for the coal reserves of type 1 above are, respectively made 3/4 and 1/2 of the standard coal seam case, as shown in the accompanying table.

In the case of the lignite coal quality (F) in the Table below the mining depth limits (successive decrease in mining depth limit according to coal thickness) are, as shown in the Table, respectively made 1/2 of the values above subbituminous coal quality (E).

Depth limiting criteria for various coal thickness classes			
Coal quality	Coal thickness Class	Mining depth limit (below mine entrance level), m.	
		Verified, estimated, predicted coal reserves type 1	Verified, estimated, predicted coal reserves type 2
Anthracite (A) Bituminous coal (B, C) Subbituminous coal (D, E)	Class 1	600	1200
	Class 2	450	900
	Class 3	300	600
Lignite (F)	Class 1	300	600
	Class 2	230	450

Note: 1 place is based on the method of counting 0.5 as a unit and discarding 0.4 and below. (JIS M 1002 - 1978).

depth limiting criteria - NCB, UK, procedure: From the assessment will be excluded areas of coal in which a seam is over 1200 m deep. (NCB, UK, Procedure for the Assessment of Reserves, 1972).

depth limiting criteria - NSW, Australia: None specified. Overburden limits imposed by the calculator for both open cut and underground reserves should be stated. (NSW Code for Coal Reserves, 1979).

depth limiting criteria - Ohai coalfield, NZ: No coal beneath more than 1500 ft of cover is included in an estimate area, and no coal with more than 1000 ft of cover is classified as **measured** or **indicated**.
Evidence in New Zealand suggests that the depth to which a coal can be worked is dependent mainly on the hardness of the country rocks, which can conveniently be considered in terms of the rank of the coal with them, and on thickness of seam and nature of roof and floor.
All coal between 1000 ft and 1500 ft in depth is classified as **inferred** and placed in a category separate from inferred coal with less than 1000 ft of cover. (NZ geol. Surv. Bull. No. 51, 1964).

depth limiting criteria - Queensland, Australia: None specified, but see spatial distribution of assessed coals for implication that limiting depth exceeds 600 m. (Queensland Coal Reserves Classification, 1977).

depth limiting criteria - RAG, FRG: **geological reserves** are to be determined to a depth of 1500 m below Ordnance Datum (-1500 m NN). (Ruhrkohle A.G. Guidelines, 1970).

depth limiting criteria - Reefton coalfield, NZ: No coal below an arbitary maximum thickness of cover is included. The limit of cover will differ from one coalfield to another, depending on the hardness of the rock, the thickness of the coal, the dip of the coal measures, the type of roof and floor of the seams. Suggested limits, where none based on experience is available, are 2500 ft in low and medium volatile bituminous, 2000 ft in high volatile bituminous A, 1500 ft in high volatile bituminous B, 900 ft in high volatile bituminous C and subbituminous, and 400 ft in lignite coalfields. (The groupings used here are those accepted by the American Society for Testing Materials). Where several feet of coal can be left in roof and floor in a thick seam, mining may be possible to greater depths in the lower rank coalfields, but this requires to be tested by experience. (NZ geol. Surv. Bull. No. 56, 1957).

depth limiting criteria - Taiwan: The lower limit of mineable coal for different reserve categories is usually defined by the thickness range and the dip of the coal bed. The mineable depth of a coal bed for the three classes of reserves is given below:

Thickness Class	Measured	Indicated	Inferred
A	0 - 1000 m	1000 - 2000 m	beyond 2000 m
B	0 - 800 m	800 - 1600 m	beyond 1600 m
C	0 - 600 m	600 - 1200 m	beyond 1200 m

For coal beds dipping less than 40°, the mineable depth of each class of reserves is measured along the inclination of the coal bed and perpendicular to the strike. For coal beds dipping more than 40°, the mineable depth of each class of reserves shall be reasonably corrected and reduced according to the magnitude of the dip and thickness range of the coal bed. In any developed coalfield, the lower limit of the mineable depth of a coal bed varies with the depth of the mine workings. Additional depth should be added to the stipulated depth range of the **measured reserves** and is in amount equal to the depth of the mined-out areas. However, as the mine workings go down beyond a depth of 500 m along the inclination of the coal beds, this added depth is reduced pro-

D

portionally. In an undeveloped coalfield where known data are not sufficient to calculate the measured coal reserves, the mineable depth of the indicated or inferred coal for that coal field should be calculated in the same order of magnitude as that for the measured coal. (Bull. geol. Surv. Taiwan, No. 10, 1959).

depth limiting criteria - USA: These apply only to those coal bodies that are or will be economically extractable by underground, surface, and/or in situ methods. All coal deeper than 6000 ft (1800 m) is excluded from Department of the Interior estimates from January 1, 1975. Deeper coals will be considered at a later date. The reserve base includes anthracite, bituminous and subbituminous coals to a depth limit of 1000 ft (300 m) and lignite to a depth of 120 ft (40 m). However, identified resources and the reserve base include coals that are deeper than the general criteria permit, but that are being mined or are judged to be mineable commercially at this time. (US geol. Surv. Bull. 1450-B, 1976).

depth method - : See reserve calculation method - Japanese standard procedure.

Dilatation (in Audibert-Arnu test,) - FRG standard definition: The percentage change in length in relation to its original length shown by a coal pencil between the end of the contraction process and solidification. (DIN 51 739, 1976, para. 3).

Dilatationsverlauf - German term: See course of dilatation - (in Audibert-Arnu test) - FRG standard definition.

dip (Einfallen; pendage) - CEC Analysis of Terms: This is the maximum angle between the surface of a stratum and the horizontal, given in gons (100 gons = 90°) or grades. (CEC Assessment of Coal Reserves, 1980).

dip - FRG standard definition: For technical, economic, and statistical purposes deposits are subdivided according to their average dip into groups. The following table gives the ranges of the groups and their description.

Average Dip		Description
Gon	Degrees	
0 to 20	0 to 18	shallow
Over 20 to 40	Over 18 to 36	gently inclined
Over 40 to 60	Over 36 to 54	steeply inclined
Over 60 to 100	Over 54 to 90	steep

(DIN 21 952, 1958).

dip limiting criteria - Reefton coalfield, NZ: No seam dipping more than 40° and less than 70° is included. Vertical or near-vertical seams have been successfully worked in some districts, but other seams dipping more steeply than 25° are difficult to mine. Some mines, large and small, have, however, extended their workings into areas with dips up to 40°, which is used as the limit in compiling estimates. (NZ geol. Surv. Bull. No. 56, 1957).

dirt - NCB, UK, definition: Dirt is defined as material with ash exceeding 40.0%. (NCB, UK, Procedure for the Assessment of Reserves, 1972).

dirt band (Bergemittel, Mittel; intercalations stériles) - CEC Analysis of

Terms: Intercalated bands of surrounding rock and inclusions in usable seams, parts of seams or layers. (CEC Assessment of Coal Reserves, 1980).

dirt band or **shale band** - ISO definition: A layer of mineral matter lying parallel to the bedding plane in a seam of coal and thicker than a **parting**. (ISO/R 1213/II - 1971, 1.21).

dirt bands - Austrian guidelines: In seam thickness evaluation, dirt bands and **partings** of a thickness below 10 cm shall not be considered; known dirt bands and partings with a thickness of 10 cm and above shall be deducted from the total thickness. (Austrian Guidelines for Coal Deposit Assessment, 1972).

dirt bands - Indian standard procedure: Partings greater than 5 cm (2 in) in thickness and the burnt out portions of coal and **jhama** shall be excluded while taking thickness of the seams for the purpose of calculation. (Indian Standard Procedure for Coal Reserve Estimation, 1977).

dirt content - RAG, FRG, guidelines: The calculation of **geological reserves** is to include only reserves in seams or beds where the proportion of dirt including dirt which will fall in as a result of geological factors is no more than 50% by weight of the total. However, seams with fairly high dirt content which are currently worked are also to be included. The dirt content is to be calculated from the accompanying graph. (Ruhrkohle A.G. Guidelines, 1970).

disposition des gisements - French term: See **inclinations of strata** - CEC Analysis of Terms.

division - : See **sample division** - US standard definition.

D:K (Decke:Kohle) - : See **geological thickness overburden ratio** - FRG standard definition.

dmmf - Abbreviation: See **dry, mineral-matter-free**.

dry (Wasserfrei) - German standard definition: Fuel which has been dried to constant weight at (106 ± 2)°C is known as dry. (DIN 51 700, 1967).

dry - : See also **dry, ash-free; dry basis; dry, mineral-matter-free**.

dry, ash-free basis - US standard definition: Data calculated to a theoretical base of no moisture or ash associated with the sample. Numerical values as established by Method D 3173 and Method D 3174 are used for converting the as-determined data to a moisture and ash-free basis. (ASTM D 3180 - 74).

dry, ash-free (daf) - : See also **dry basis; dry, mineral-matter-free**.

dry, ash-free (daf)[wasser-und faschefrei (waf)] - FRG standard definition: Fuel which has been dried to constant weight at (106 ± 2)°C is known as **dry**. The dry, ash-free condition is a theoretical concept. the weight of ash remaining after combustion at (815 ± 10)°C is subtracted from the weight of dry fuel. Dry, ash-free matter does not correspond to **inerts-free** fuel, since the mineral inerts proportion generally changes its weight when the fuel is incinerated. More often than not, weight decreasing processes predominate so that analytical data related to dry, ash-free coal are mostly lower than corresponding data for the organic part, because the reference quantity for the proportions is larger. Generally, however, the differences are so small that relation to dry, ash-free coal suffices for practical purposes (DIN 51 700, 1967, para. 5.2).

D

dry basis - US standard definition: Dry basis - data calculated to a theoretical base of no moisture associated with the sample. The numerical value as established by Method D 3173 is used for converting the as-determined data to a dry basis. (ANSI/ASTM D 3180 - 74, para. 3.3).

dry-cleaned coal - UK standard definition: Coal from which impurities have been removed mechanically without the use of liquid media. (BS 3323 : 1978).

dry, mineral-matter-free basis (dmmf basis) - Australian standard definition: The hypothetical condition in which coal or coke is calculated to be free of both moisture and mineral matter. (AS K149 - 1966).

dry, mineral-matter-free (wasser-und mineralstoffrei) - FRG standard definition: Dry, mineral-freecoal is the inerts-free proportion, i.e. the organic fuel. In order to calculate data on a dry, mineral-free basis, the mineral content must be reported. (DIN 51 700, 1967, para. 5.2).

dry, mineral-matter-free (dmmf) - NCB, UK, classification: For calculation to the dry, mineral-matter-free basis, the following equation is applied in the case of volatile matter:-

$$\text{V.M. dmmf} = \frac{(\text{V.M.} - \text{correction}) \times 100}{100 - M - M.M.}$$

where:
V.M. dmmf = volatile matter on the dry, mineral-matter-free basis.
V.M. = volatile matter on the air-dried (or as analysed) basis.
Correction = as below, constituents being on the air-dried (or as analysed) basis.
M = moisture on the air-dried (or as analysed) basis.
M.M. = mineral matter on the air-dried (or as analysed) basis.

Calculation of correction: The correction is calculated by one of of the following equations:

In cases where both pyritic sulphur and chlorine have been determined:
Correction = 0.13ash + 0.2S(pyritic) + 0.7CO2 + 0.7Cl - 0.2;

In cases where chlorine but not pyritic sulphur has been determined:
Correction = 0.13ash + 0.2S(total) + 0.7CO2 + 0.7Cl - 0.32;

In cases where neither pyritic sulphur nor chlorine has been determined:
Correction = 0.13ash + 0.2S(total) + 0.7CO2 - 0.12;

Calculation of Mineral Matter:
Whenever possible mineral matter is calculated by the following equation:-
M.M. = 1.13ash + 0.5S(pyritic) + 0.8CO2 - 2.8S(ash) + 2.8S(sulphate) + 0.3Cl.

In the absence of sufficient analytical data to calculate mineral matter by this formula, the following equation is used:-
M.M. = 1.10ash + 0.53S(total) + 0.74CO2 - 0.36.
(NCB Coal Classification System, 1964).

dry mineral-matter-free - US standard procedure: For classification of coal according to rank, fixed carbon and volatile matter shall be calculated to the mineral-matter-free basis in accordance with either the Parr formulas, Eqs. 1

and 2, or the approximation formulas, Eqs. 4 and 5, that follow. In the case of litigation, use the appropriate Parr formula.
Calculation to MM-free basis: Parr formulas:
Dry MM-free FC = (FC - 0.15S) / [100 - (M + 1.08A + 0.55S)] x 100   (1)
Dry, MM-free VM = 100 - Dry, MM-free FC   (2)
Note - The above formula for fixed carbon is derived from the Parr Formula for volatile matter.
Approximation formulas:
Dry, MM-free FC = FC / '100 - (M + 1.1A + 0.1S)' x 100   (4)
Dry, MM-free VM = 100 - Dry, MM-free FC   (5)
where:-
MM = mineral matter,
FC = percentage of fixed carbon,
VM = percentage of volatile matter,
M = percentage of moisture,
A = percentage of ash,
S = percentage of sulphur.
Above quantities are all on the **inherent moisture** basis. This basis refers to coal containing its natural inherent or bed moisture but not including water adhering to the surface of the coal. See Method D 3180, 5.1.3, for calculating analyses to this basis. (ANSI/ASTM D 388, - 77, para. 8).

dry steam coal - UK standard definition: Coal of higher volatile content than **anthracite**, falling within the general category of '**low volatile**' coal. Note: It corresponds to NCB code number 201. (BS 3323 : 1978).

dull coal - UK standard definition: Coal of dull appearance, which may be **durain**, or the dullness may be imparted by a high content of mineral matter. (BS 3323 : 1978).

durain - UK standard definition: A strong dull coal with an uneven fracture. (BS 3323 : 1978).

E

economic (abbauwürdig) - Austrian guidelines: Classified in this category are seams with a lower and upper limit of thickness which can be mined in accordance with current practice and the current equipment level of the enterprise in question. Protective roadway pillars are taken into account by means of an empirically determined percentage. (For new deposits allowance has to be made for stoping method and equipment levels which also have to be included in the calculation of costs for the assessment of economic feasibility). (Austrian Guidelines for Coal Deposit Assessment, 1972).

economic (bauwürdig) - FRG standard definition: In respect of seams lying outside developed areas, refers to those which may be economically worked at present and forseeable levels of mining technology. (DIN 21 941, 1953).

economic mineability (Abbauwürdigkeit) - Austrian guidelines: Assessment of economic mineability relates to a definite point in time and is, therefore, variable with time. Resource groups A and B, but not C1 and C2, are subdivided into economic, subeconomic, and uneconomic. (Austrian Guidelines for Coal Deposit Assessment, 1972).

economically recoverable reserves (wirtschaftlich bauwürdige Vorräte; réserves économiquement exploitables) - CEC Analysis of Terms: To define economically recoverable reserve is a matter of full ranged importance. For short time intervals and on smallest scales they are to be defined as that small part of large geological reserves for which the proceeds obtainable on a free market are known to be most likely to exceed the expenditures. For long time intervals and on large scales they are the part of the geological reserves, for which the benefits of their recovery are known to be greater than their disadvantages. Consequently the last named reserves are always of a larger amount than the first named. (CEC Assessment of Coal Reserves, 1980).

Einfallen - German term: See dip - CEC Analysis of Terms.

equilibrium moisture - : See inherent moisture (in coal) - US standard definition.

equilibrium moisture base - US standard definition: Data calculated to the moisture level established as the equilibrium moisture. Numerical values as established by Method D 1412 are used for the calculation. (ANSI/ASTM D 3180 - 74).

Erforschungsgrad - German term: See degree of investigation - CEC Analysis of Terms.

Erkundungsgrad - German term: See degree of prospecting - CEC Analysis of Terms.

Erweichungstemperatur - German term: See softening temperature - FRG standard definition.

Esskohle - German term for steam coal: See low volatile coal - RAG, FRG, definition.

estimated coal reserves - Japanese standard classification: This is one of three categories into which coal reserves are generally divided, the others being verified coal reserves and predicted coal reserves. All these are further divided into types 1 and 2 depending on bed depth. (JIS M 1002 - 1978).

estimated coal reserves type 1 - Japanese standard definition: Coal seams within geological structures from information other than outcrops, mines, and drilling in areas adjoining verified coal reserves type 1 (B) are coal reserves in estimated areas, are within the present mining depth limit, and are of low assurance in comparison with verified coal reserves type 1 (B). (JIS M 1002 - 1978).

estimated coal reserves type 2 - Japanese standard definition: In terms of assurance these belong to estimated coal reserves type 1 but the bed depth is beyond the present mining depth limit, being within the future mining depth limit. (JIS M 1002 - 1978).

exinite - UK standard definition: The group of macerals derived from plant cell secretions, cuticles and outer membrances of spores and pollen grains. It is normally the minor maceral component of coal. (BS 3323 : 1978).

exinite (or liptinite) - US standard definition: A group of macerals composed of alginite, cutinite, resinite, and sporinite. (ANSI/ASTM D 2796 - 77).

extractable coal - Rep. South Africa, definition: that portion of the mineable coal in situ which is extractable in the prevailing or slightly less rigorous condition of mineable coal in situ Table 5.31 of ref. specifies conditions in terms of minimum size, thickness and depth range and extraction %, for surface and underground extraction; by types of coal. See also - mineable coal in situ. (Dept. Mines, Rept. Coal Res. SA 1975).

extraction rate - : See recovery factor - Ohai and Reefton coalfield, NZ.

extraneous ash - UK standard definition: Ash arising from mineral matter associated with, but not inherent in, coal. (BS 3323 : 1978).

F

facteur de pertes - French term: See **loss factor** - CEC Analysis of Terms.

facteur de rendement - French term: See **recovery factor** - CEC Analysis of Terms.

fat coal - UK standard definition: Coking coal with a medium **volatile matter** content. Note: This term is not used in the United Kingdom but is in general use in some other countries. (BS 3323 : 1978).

fat (coal) - Literal translation of term in Belgian definitions:
1/2 vet, 1/2 gras   : from 14% to less than 18% V.M.
3/4 vet, 3/4 gras   : from 18% to less than 20% V.M.
Vet A, gras A   : from 20% to less than 28% V.M.
To a basis of dry coal containing 5% of ash. ("Moniteur Belge-Belgisch Staatsblad", 19 August 1971, Page 9664).

fat coal (Fettkohle) - RAG, FRG, definition: V.M. (daf) by weight: 18% to 30%; crucible coke strongly caked, solid. (Ruhrkohlen-Handbuch, 1969, para. 1.1.1.1 Table).

fat (fett, gras, vet) - Literal translation of German, French and Flemish terms: See also **bituminous coal; medium volatile coal; high volatile coal.** "Fat" applied to coal and bituminous coals are often taken as synonymous terms, but "fat" implies caking ability whereas "bituminous" does not.

faults - : See **tectonic disturbance** - Austrian guidelines and Indian standard procedure.

fett - German term meaning fat : See **fat coal** - RAG, FRG, definition.

Fettkohle - : See **fat coal** - RAG, FRG, definition.

fine sample - : See **analysis sample** - FRG standard definition.

fixed carbon - Indian standard definition: Obtained by subtracting from 100 the sum of the percentages by weight of **moisture, volatile matter** and **ash**. [IS: 1350 (Part I) 1969].

fixed carbon - UK standard definition: A calculated figure obtained by subtracting the sum of the percentages of **moisture, volatile matter** and ash from 100. (BS 3323 : 1978).

fixed carbon - US standard definition: In the case of coal, coke, and bituminous materials, the solid residue other than ash, obtained by destructive distillation determined by definite prescribed methods.
Note 1 - It is made up principally of carbon, but may contain appreciable amounts of sulphur, hydrogen, nitrogen, and oxygen.

Note 2 - In the case of coal and coke, the methods employed shall be those prescribed in ASTM Method D 3172, Proximate Analysis of Coal and Coke. (ANSI/ASTM D 121 - 76).

flüchtigen Bestandteilen - German term: See **volatile matter** - FRG standard definition.

free-burning coal - UK standard definition: Coal that does not cake during combustion in a fuel bed and which has a high volatile matter content. (BS 3323 : 1978).

free impurity - US standard definition: The impurities in a coal that exist as individual discrete particles that are not a structural part of the coal and that can be separated from it by coal preparation methods. (ANSI/ASTM 2234 - 76).

free moisture - ISO definition: The moisture which is lost by the coal sample in attaining approximate equilibrium with the air to which it is exposed (see ISO Recommendation R 1988, Sampling of Hard Coal). (ISO/R 1213/II - 1971, 3.07).

free moisture - UK standard definition: The moisture that is lost by the coal sample in attaining equilibrium with ambient laboratory conditions. (BS 3323 : 1978).

free moisture - US standard definition: Free moisture in coal: That portion of **total moisture** in coal (determined in accordance with Method D 3302) that is in excess of **inherent moisture** in coal (determined in accordance with ASTM Method D 1412, Test for the **equilibrium moisture** of coal at 96% to 97% relative humidity and 30°C). It is not to be equated with the weight loss upon air drying. Free moisture is sometimes referred to as **surface** moisture in connection with coal. (ANSI/ASTM D 121 - 76).

free moisture (grobe Feuchtigkeit) - FRG standard definition: Crude moisture or surface moisture is that water which evaporates when the fuels lie in the air at room temperature. (DIN 51 718, 1978, para. 2).

fusain - UK standard definition: A soft, friable form of coal occurring in seams as thin bands or lenses and often showing a fibrous structure and silky lustre. (BS 3323 : 1978).

fusain - US standard definition: Coal layers composed of chips and other fragments in which the original form of plant tissue structure is preserved; commonly has fibrous texture with a very dull luster. (ANSI/ASTM D 2796 - 77).

fusinite - US standard definition: The maceral distinguished by the well-preserved original form of plant cell wall structure, intact or broken, with open or mineral-filled cell lumens (cavities) and by having a reflectance (except in meta-anthracite), well above that of associated vitrinite. (ANSI/ASTM D 2796 - 77).

G

gas coal - RAG, FRG, guidelines:

   Gaskohle   : V.M. (daf) by weight: 28% to 33%; crucible coke caked and fissured.
   Gasflammkohle : V.M. (daf) by weight: 33% to 40%; crucible coke sintered, in part lightly caked.
   (Ruhrkohlen-Handbuch, 1969, para. 1.1.1.1 Table).

Gasflammkohle - German term applicable to a high volatile coal : See gas coal high volatile coal - RAG, FRG, guidelines.

Gaskohle - German term: See gas coal - RAG, FRG, guidelines.

gazowo-koksowy wegiel - Polish term for a class of coals embracing gas coal and coking coal: - See high volatile coals - Polish standard classification.

gazowo-plomienny wegiel - Polish term for long-flame gas coal: See high volatile coals - Polish standard classification.

gazowy-wegiel - Polish term for gas coal: See high volatile coals - Polish standard classification.

general analysis - ISO definition: (See also proximate analysis, ultimate analysis. This term, frequently used in English publications, means the determination of the chemical and physical characteristics of coal, other than the determination of moisture. (ISO/R 1213/II - 1971, 3.20).

geological reserves - : See coal in place (geological reserves) - CEC Analysis of Terms.

geological reserves - : See geological resources - RAG, FRG, guidelines.

geological resources - RAG, FRG, guidelines: The total coal resources determined in accordance with these guidelines are termed the geological resources and are subdivided into tectonic areas by seams or beds, taking into account the standard seam designation. (Ruhrkohle A. G. Guidelines, 1970).

geological resources (risorse geologiche) - Italian definition: "These are all those deposits for which the degree of geologic knowledge is sufficient to define them. These may be considered "Coal Reserves". (ENI-ENEL-FINSIDER Classification Systems for Coals, 1978).

geological thickness (geologische Mächtigkeit; puissance géologique) - CEC Analysis of Terms: The distance, normal to the stratification, between the stratigraphic limiting planes of a stratum. (CEC Assessment of Coal Reserves, 1980).

G

geological thickness overburden ratio (D:K Decke:Kohle) - FRG standard definition: For brown coal, is calculated from the thicknesses of the covering rock and coal in place as revealed by boreholes or surface mine development. It is the total cover thickness (cover thickness to the roof of the top seam plus the total of all seam interburden thicknesses) in relation to the total of all coal thicknesses. (DIN 21 942, 1961).

geologically inferred - Austrian guidelines: See **resource group C2**. (Austrian Guidelines for Coal Deposit Assessment, 1972).

geologische Mächtigkeit - German term: See **geologial thickness** - CEC Analysis of Terms.

geologischer Vorrat - German term: See **coal in place (geological reserves)** - CEC Analysis of Terms.

gering bituminöse Kohle - German term: See **low volatile coals** - FRG standard definition.

Gesamtwasser - German term: See **total moisture** - FRG standard definition.

gewinnbare Vorräte - German term: See **recoverable resources** - FRG standard definition.

gewogene Mächtigkeit - German term: See **weighted thickness** - CEC Analysis of Terms.

Grad der tektonischen Beanspruchung - German term: See **degree of tectonic stress** - CEC Analysis of Terms.

grade 1 reserves - CdF, France, classification: From the standpoint of technical and economic feasibility in the light of the present situation, these are promising reserves in economic terms which are likely to offer the requisite output pattern. The thickness (or mined section) of a seam to be included in the calculation of grade 1 reserves can be less than the total average thickness (e.g. part of a seam may be economically workable even though the whole is not). The following are some of the factors to be taken into account when classifying a panel under this heading:
i) seam structure and quality of roof and floor,
ii) known mechanisation potential at the time of classification,
iii) drivage and equipment required to work the panel, taking into account its reserve,
iv) the possible upgrading of the products and their saleability.
Grade 1 is a subclassification of categories "a", "b", and "c". (CdF Reserves Classification, 1972 Model).

grade 2 reserves - CdF, France, classification: A subclassification of category "a", "b" and "c" reserves, these are reserves which a mine is technically capable of working, but which are judged inadequate in economic terms (output too low for an economic return, too difficult to market the coal, capital outlay too great, etc). Of course, reserves which a mine is technically incapable of working are to be disregarded. (CdF Reserves Classification, 1972 Model).

gram calorie (cal) - Indian standard definition: "Amount of heat required to raise the temperature of one gram of water from 14.5°C to 15.5°C, used as an international standard for calorimetry. It is equal to 4.1855 Joules and it differs from the mean calorie by only one part in 1000, which is not significant enough to require correction for practical purposes." [ISO: 1350 (Part II)

G

- 1970].

gras - French term meaning fat: See **fat coal** - Belgian standard definition.

Gray-King coke type - Australian standard definition: A letter which defines, by reference to a series of standard profiles, the size and texture of the coke residue produced when a standard weight of coal is heated in a retort tube to 600°C under standard conditions. (AS K184-1969, para. 3.3).

Gray King coke type - UK standard definition: The letter and/or number that defines, by reference to a series of standard profiles, the size and texture of the coke residue produced when a standard mass of coal is heated in a retort tube to 600°C under standard conditions. For certain coals, numbered subscripts applied to the letter G represent the proportion of electrode carbon or specially prepared anthracite required in admixture to produce a coke residue of profile G. (BS 3323 : 1978).

Gray-King coke type - UN ECE classification: The purpose of the method is to assess the caking properties of a coal or a blend of coals by carbonizing in a laboratory assay under standard conditions... The coke residue from the carbonization of finely ground coal at 600°C is classified by comparison with a series of described coke types. ("International Classification of Hard Coal by Type, 1956, app. IV).

grobe Feuchtigkeit - German term: See **free moisture** - FRG standard definition.

gross calorific value - Australian standard definition: The number of heat units measured as being liberated per unit quantity of fuel burned in oxygen in a bomb under standard conditions in such a way that the material after combustion (suitable corrections having been made) consists of gaseous oxygen, carbon dioxide and nitrogen, liquid water in equilibrium with its vapour and saturated with carbon dioxide, and **ash**. (AS K184, 1969, para. 3.7).

gross calorific value - Indian standard definition: Number of heat units liberated when a unit mass of the fuel is burnt at constant volume in oxygen saturated with water vapour, the original material and final products being at approximately 25°C. The residual products are taken as carbon dioxide, sulphur , dioxide nitrogen and water; the residual water other than that originally present as vapour, being in the liquid state.

Note 1 - The gross calorific value at constant volume is the one usually used in coal technology. It is assumed that all the heat produced is available, including the heat of condensation of any steam, resulting from the combustion of hydrogen of the fuel, to water at room temperature.

Note 2 - To convert gross calorific value to net calorific value employ the formula: $Nc = Gc - 53H$
where:
    Nc = net calorific value in kcal/kg,
    Gc = gross calorific value in kcal/kg, and
    H = percentage of hydrogen present in the coal sample,
        including hydrogen of moisture and of water of
        constitution.
(IS: 1350 (Part II) - 1970).

gross calorific value - ISO definition: The number of heat units measured as being liberated when unit mass of coal is burnt in oxygen saturated with water vapour in a bomb under standarized conditions (as defined in ISO Recommendation

R 1928, Solid Mineral Fuels - Determination, and Calculation of Net Calorific Value), the residual materials being taken as gaseous oxygen, carbon dioxide, sulphur dioxide and nitrogen, liquid water in equilibrium with its vapour and saturated with carbon dioxide, and ash. (ISO/R 1213/II - 1971, 3.12).

gross calorific value at constant volume - UK standard definition: The number of heat units liberated when unit mass of fuel is burned in oxygen in a bomb calorimeter under standard conditions. The materials present after combustion (suitable corrections having been made) are assumed to be oxygen, carbon dioxide, nitrogen, liquid water in equilibrium with its vapour and saturated with carbon dioxide, hydrochloric acid in solution and solid ash. (BS 3323 : 1978).

gross calorific value (Brennwert) - FRG standard definition: The gross calorific value H is the ratio of the amount of heat generated by the complete combustion of a solid or liquid fuel to the weight of fuel when: a) the temperature of the fuel before combustion and that of the products of combustion is 25°C; b) the moisture present in the fuel prior to combustion and the moisture formed during the combustion of the hydrogen-bearing compounds in the fuel remains in liquid form; c) the products of the combustion of carbon and sulphur remain as carbon dioxide and sulphur dioxide in a gaseous state; and d) the nitrogen has not oxidised. (DIN 51 900, Part I, 1977, para. 32).

gross calorific value (gross heat of combustion), H(gs) - US standard definition: In the case of solid fuels and liquid fuels of low volatility, the heat produced by combustion of unit quantity, at constant volume, in an oxygen bomb calorimeter under specified conditions. (ANSI/ASTM D 121 - 76).

gross (specific) calorific value [Brennwert (spezifischer); coefficient calorifique (spécifique)] - CEC Analysis of Terms: This is the amount of heat of reaction (enthalpy in the case of combustion under constant pressure) liberated by a given mass of a solid or liquid fuel on complete combustion. It is determined in accordance with standards which lay down, among other things, the physical and chemical conditions. Internationally ISO/R 1928-1971 applies. (CEC Assessment of Coal Reserves, 1980).

gross heat of combustion - : See **gross calorific value** - US standard definition.

gross moisture sample - US standard definition: A sample representing one lot of coal and composed of a number of increments on which neither reduction nor division has been performed or a subsample for moisture testing that is taken in accordance with Section 10 of Method D 2234. (ASTM D 3180 - 74).

gross sample - ISO definition: A sample formed when all the increments collected from a consignment (batch or unit) are combined for reduction to a laboratory sample: where two or more samples are formed from interleaved increments, these samples are designated duplicate samples or replicate samples as the case may be. (ISO/R 1213/II - 1971, 2.15).

gross sample - US standard definition: A sample representing one lot of coal and composed of a number of increments on which neither reduction nor division has been performed. (ANSI/ASTM 2234 - 76).

gross thickness (Bruttomächtigkeit; puissance brute) - CEC Analysis of Terms: Total thickness of recoverable strata including intermediate rock and dirt bands in a deposit or part of a deposit. (CEC Assessment of Coal Reserves, 1980).

H

hard coal - Australian standard definition: Coal having or exceeding a **gross calorific value** of 6470 kcal/kg on a **dry, ash-free (daf)** basis. (AS K184 - 1969, para. 3.1).

hard coal - UK standard definition: All coal higher rank than brown coal and lignite. Note: In the USA the term is restricted to anthracite. (BS 3323 : 1978).

hard coal - UN ECE definition: Hard coal is defined for the purpose of internation classification as coal with a **gross calorific value** over 5700 kcal/kg on the **moist, ash-free** basis. (International Classification of Hard Coals by Type, 1956, ch. 2).

hard coal (carboni) - Italian definition: Having a **gross calorific value** greater than 5700 kcal/kg on the basis of **mineral-matter-free** coal in equilibrium with air at 30°C and 96% relative humidity, they are subdivided on the basis of **volatile matter (V.M.)** content on a **dry, ash-free (d.a.f.)** basis, a parameter characteristic of the rank. A more precise parameter for characterizing the rank is the index of reflectivity of the **vitrinite**; this, however, is a more complex measurement.
See **antraciti, litantraci basso volatili, litantraci medio volatili, litantraci alto volatili A,** and **litantraci alto volatili B** for the subdivisions of hard coal. (ENI-ENEL-FINSIDER Classification Systems for Coals, 1978).

heat-altered coal - UK standard definition: Coal that has been affected in situ by the effects of igneous intrusion. (BS 3323 : 1978).

heat of combustion - equivalent to calorific value: See **gross calorific value; net calorific value.**

Heizwert - German term: See **net calorific value** - FRG standard definition.

Heizwert (spezifischer) - German term: See **net (specific) calorific value** - CEC Analysis of Terms.

high volatile A bituminous coal - : See **high volatile coal** - US standard definition.

high volatile a bituminous coal (litantraci alto volatili A) - Italian classification. These coals have a volatile matter content from 28% to 42% on a dry, ash-free basis within the general classification of **hard coals** and an index of reflectivity of **vitrinite** less than 1.1%. They have high coking power when their dilatation is positive, more than 150%, with a free swelling index from 7 to 9; medium coking power with a positive dilatation from 80% to 150% with a free swelling index from 4 to 6.5; low coking power with a positive dilatation from 30% to 80% and with a free swelling index from 1.5 to 3.5; and no coking power when in the dilatometer test there is either only contraction or dilatation from negative to 30% positive, with a free swelling index from 0 to 1. (ENI-ENEL-FINSIDER Classification Systems for Coals, 1978).

high volatile B bituminous coal - : See high volatile coal - US standard definition.

high volatile b bituminous coal (litantraci also volatili B) -Italian classification. These coals have a volatile matter content above 42% on a dry, ash-free basis, within the general classification of hard coals. (ENI-ENEL-FINSIDER Classification Systems for Coals, 1978).

high volatile C bituminous coal - : See high volatile coal - US standard definition.

high volatile coal - Indian standard classification: See **subbituminous** and **bituminous coals** - Indian standard classification.

high volatile coal - UK standard definition: Coal with a volatile matter content (DMMF) greater than 32%. Note: It corresponds to NCB rank codes 400 to 900. (BS 3323 : 1978).

high volatile coal - US standard definition:

Group	FC % (dmmf) Less Than	VM % (dmmf) Greater Than	(C.V. Btu/lb) (Moist mmf) Equal or Greater Than	Less Than	Agglomerating Character
High volatile A bituminous coal	69	31	14000	-	
High volatile B bituminous coal	-	-	13000	14000	commonly agglomerating
High volatile C bituminous coal	-	-	11500 10500	13000 11500	agglomerating

(ANSI/ASTM D 388 - 77, Table 1).

high volatile coal (Gasflammkohle) - : See gas coal - RAG, FRG, guidelines.

high volatile coals - : See also bituminous coal, medium volatile coals and subbituminous coals.

H

high volatile coals - FRG standard definition:

Name (indication)	International Class (first Index no.)	Volatile Matter Content in dry, ash-free matter	Calorific Value of air-dry ash-free matter kcal/kg
hoch bituminöse Kohle (high volatile) bituminous coal)	5	> 28 to 33	-
	6	> 33 (33 to 41)	(> 32400)
	7	> 33 (33 to 44)	(32400 to > 30100)
	8	> 33 (35 to 50)	(30100 to > 25500)
	9	> 33 (42 to 50)	(25500 to > 23900)

(DIN 23 003, 1976, Table 1).

high volatile coals - Italian definition:
  litantrici alto volatili A : V.M. (daf) 28% to 42%
  litantrici alto volatili B : V.M. (daf) over 42%
      both qualities : gross calorific value moist, ash-free > 5700 kcal/kg; reflectance of vitrinite less than 1.1%.

(ENI-ENEL-FINSIDER Classification Systems for Coals, 1978).

high volatile coals - NCB, UK, classification:

Coal Rank Code Main Class(es)	Class	Volatile Matter (d.m.m.f) (per cent.)	Gray-King Coke Type	General Description
400 to 900:-		Over 32.0	A-G9 and over	High-volatile coals
400		Over 32.0	G9 and over	)High-volatile, very
	401	32.1-36.0	) G9 and	)strongly caking
	402	Over 36.0	) over	)coals
500		Over 32.0	G5-G8	)High-volatile,
	501	32.1-36.0	) G5	)strongly caking
	502	Over 36.0	) -G8	)coals
600		Over 32.0	G1-G4	)High-volatile,
	601	32.1-36.0	) G1	)medium-caking
	602	Over 36.0	) -G4	)coals
700		Over 32.0	E-G	)High-volatile,
	701	32.1-36.0	) E	)weakly caking
	702	Over 36.0	) -G	)coals
800		Over 32.0	C-D	)High-volatile,
	801	32.1-36.0	) C	)very weakly
	802	Over 36.0	) -D	)caking coals
900		Over 32.0	A-B	)High-volatile,
	901	32.1-36.0	) A	)non-caking
	902	Over 36.0	) -B	)coals

(N.C.B. Coal Classification System, 1964).

H

high volatile coals - Polish standard classification:

Coal Type		Classification criteria			General Technical characteristics
Name	Code Number	Volative Matter (daf) %	Caking Properties Roga Index	Heat of u-c combustion (daf) kcal/kg	
Long-flame coal Wegiel plomienny	31.1	Over 28	5 or below	7400 or below	High volatile matter content, usually 35-44%; practically no caking properties
	31.2			over 7400	
Long-flame gas coal	32.1	Over 28	Over 5 up to 20	Not Standard-ised	High volatile matter content; usually 33-40%; very poor caking properties
Wegiel gazowo-plomienny	32.2		Over 20 up to 45	Not Standard-ised	High volatile matter content, usually 33-40%; medium caking properties
Gas coal Wegiel gazowy	33	Over 28	Over 45 up to 55	Not Standard-ised	High volatile matter content, good caking properties
Gas/Coking coal; Wegiel gaz-owo-koksowy	34	Over 28	Over 55	Not Standard-ised	High volatile matter content, good caking properties

Volatile matter determined according to Polish standard PN-71/G-04516 (Hard Coal);
Caking properties determined according to Polish standard PN-69/G-04518 (Hard Coal);
Dilatation criteria are not specified.
Heat of combustion and calorific value determined according to Polish standard PN-67/G-04513.
(PN-68/G-97002).

hvAb - Abbreviation of US term: See **high volatile A bituminous coal** - US standard definition.

hvBb - Abbreviation of US term: See **high volatile B bituminous coal** - US standard definition.

hvCb - Abbreviation of US term: See **high volatile C bituminous coal** - US standard definition.

H

hygroscopic moisture, (hygroskopische Feuchtigkeit) - FRG standard definition: Hygroscopic moisture is that moisture which evaporates additionally when air dried coals are dried at $(106 \pm 2)°C$. (DIN 51 718, 1978, para. 2).

hygroskopische Feuchtigkeit - German term: See **hygroscopic moisture** - FRG standard definition.

hypothetical resource - US official definition: "An undiscovered virgin coal resource that may be expected to exist in a known coal mining area similar to or as an extension of a known coal deposit or as a more distant deposit existing under analogous geologic conditions. In general, hypothetical resources are in the central parts of broad areas of coal fields where points of measurements are absent and evidence for thickness and existence is from distant outcrops, mine-workings, drill-holes, and wells. Exploration by geologic mapping, geophysical surveying, and drilling that confirms their existence and allows more reliable estimation of quantity, quality, and rank will permit their reclassification as **reserves, marginal reserves,** and **subeconomic resources**" - Criteria: "A tonnage estimate for this category of coal is quantitative and is based on knowledge of the geologic character, habit, and pattern of a coal bed, beds, or region. Measurements and geologic data are projected for miles from points of control. Points of measurement of coal thickness at the outcrop, in trenches, in mine workings, in drillholes, and in wells are more than 6 miles (9.6 km) apart and assumptions of coal existence and continuity are supported only by geologic evidence". (USGS 1979 Proposed Revision of Bull. 1450-B, 1976).

hypothetical resources - US official definition: "Undiscovered coal **resources** in beds that may reasonably be expected to exist in known mining districts under known geologic conditions. In general, hypothetical resources are in broad areas of coal fields where points of observation are absent and evidence is from distant outcrops, drill holes, or wells. Exploration that confirms their existence and reveals quantity and quality will permit their reclassification as a **reserve** or **identified subeconomic resource**." - The following criteria for hypothetical resources are applicable to both the **reserve** and **subeconomic resource** components: "Quantitative estimates are based on a broad knowledge of the geologic character of a coal bed or region. Measurements of coal thickness are more than 6 miles (9.6 km) apart. The assumption of continuity of a coal bed is supported only by geologic evidence". (US geol. Surv. Bull. 1450-B, 1976).

hypothetical resources (risorse ipotetiche) - Italian definition: "These are all deposits tentatively assumed to exist on the basis of regional geologic analogies, or on the basis of evidence not sufficiently conclusive to define them". (ENI-ENEL-FINSIDER Classification Systems for Coals, 1978).

I

identificados, recursos - Spanish term: See **identified resources** - Colombian official classifiation.

identified resource - US official definition: "A coal **resource** known from specific geologic evidence derived from a resource body or deposit".
- Discussion: "Such resources may be accessed and/or virgin bodies of coal which are assigned to **resource** and **reserve** categories on the basis of geologic evidence from maps, samples, drillholes, mine records, and current field work. Specific evidence must include data on the location, depth of burial, distance from points of measurement or sampling, extent, and thickness of the resource body. Evidence about quality and rank may be determined from analyses of samples collected from the resource bodies or may be inferred by projection of analytical data obtained elsewhere in the body or from adjacent bodies. An identified resource body may contain **reserves**, marginal reserves, and subeconomic resources assignable to the measured, indicated, and inferred reliability categories".
- Criteria: "A tonnage estimate for this category of resources includes all beds of **bituminous coal** and **anthracite** 14 in (35 cm) or more thick and all beds **subbituminous coal** and **lignite** 30 in (75 cm) or more thick from the surface to depths of 6000 ft (1800 m) whose existence and quantity have been determined within specified degrees of geologic assurance as measured, indicated, and inferred. The estimate also includes thinner and/or deeper beds that are currently being mined, or for which there is evidence that they could be mined economically". (USGS 1979 Proposed Revision of Bull. 1450-B, 1976).

identified resources - US official definition: "Specific bodies of coal whose location, rank, quality, and quantity are known from geologic evidence supported by engineering measurements". - Criteria: "Include beds of **bituminous coal** and **anthracite** 14 in (35 cm) or more thick and beds of **subbituminous coal** and **lignite** 30 in (75 cm) or more thick that occur at depths to 6000 ft (1800 m), and whose existence and quantity have been delineated within specified degrees of geologic assurance as **measured**, **indicated**, or **inferred**. Include also thinner and/or deeper beds that presently are being mined or for which there is evidence that they could be mined commercially". (US geol. Surv. Bull. 1450-B, 1976).

identified resources (recursos identificados) - Colombian official classification: These are specific mineralised bodies, the location, quality and quantity of which are known from geological evidence supplemented by engineering measurements. (Publ. Geol. Esp. Ingeominas, No. 3, 1979).

identified subeconomic resources - US official definition: "The part of coal **resources** that occurs in **demonstrated** and **inferred** resources that is not now mineable economically". - See also the definitions of the following terms applicable to the identified subeconomic resource component: **measured**, **indicated**, **demonstrated**, **inferred**, **rank**, and **quality or grade**.

I

- Criteria for subeconomic resources: "Include all identified resources that do not fall into the reserve category. Include in this category beds of bituminous coal and anthracite 14 in (35 cm) to 28 in (70 cm) thick and beds of subbituminous coal 30 in (75 cm) to 60 in (150 cm) thick that occur at depths to 1000 ft (300 m). Include also beds of bituminous coal and anthracite 14 in (35 cm) or more thick and beds of subbituminous coal 30 in (75 cm) or more thick that occur at depths between 1000 ft (300 m) and 6000 ft (1800 m). Include lignite beds 30 in (75 cm) or more thick that cannot be surface mined - generally those that occur at depths greater than 120 ft (40 m), and lignite beds 30 in (75 cm) to 60 in (150 cm) thick that can be surface mined. Include the currently nonrecoverable portion of the reserve base". - See also the criteria for the following terms applicable to Subeconomic Resources components: measured, indicated, inferred, hypothetical resources, and speculative resources. (US geol. Surv. Bull. 1450-B, 1976).

inclinations of strata (Lagerungsverhältnisse; disposition des gisements) - CEC Analysis of Terms:
Seams are regarded as:
       level with dips of 0 - 20 gon
       moderately inclined with dips of 20 - 40 gon
       strongly inclined with dips of 40 - 60 gon
       steep with dips of more than 60 gon
       100 gon = 90°
(CEC Assessment of Coal Reserves, 1980).

indicadas, reservas - Spanish term: See indicated (or probable) reserves - Colombian official classification.

indicated - : See resource group C1 - Austrian guidelines.

indicated - US official definition, "applicable to both the reserve and identified subeconomic resource components of the classification system: Coal for which estimates of the rank, quality, and quantity have been computed partly from sample analyses and measurements and partly from reasonable geologic projections".
- Criteria, applicable to both the reserve and subeconomic resources components of the classification system: "Resources are computed partly from specified measurements and partly from projection of visible data for a reasonable distance on the basis of geologic evidence. The points of observation are 1/2 (0.8 km) to 1 1/2 miles (2.4 km) apart. Indicated coal is projected to extend as a 1/2-mile (0.8 km) wide belt that lies more than 1/4 mile (0.4 km) from the outcrop or points of observation or measurement". (US geol. Surv. Bull. 1450-B, 1976).

indicated coal - Ohai coalfield, NZ, classification: The definition/explanation of this term is as given for the Reefton Coalfield (1957, below).

indicated coal - Reefton coalfield, NZ, classification: This is coal for which tonnage and ash content are computed partly from specific measurements, samples, or production data and partly from projection for a reasonable distance on geological evidence. The sites available for inspection, measurement, and sampling are too widely or otherwise inappropriately spaced to outline the coal completely or to establish its ash content throughout. Beyond the limits set for the measured coal, areas of indicated coal are generally limited by a position of unworkable thin coal based on detailed isopachs or an assumed rate of thinning of the seam judged from mine workings, from outcrops, or from drill holes. The rate of thinning must be estimated for each coalfield, or for individual seams within a coalfield. Examples of such rates of thinning are:

Greymouth Coalfield   1 ft per 2 chains
Reefton   Coalfield
           No. 4 seam   2 ft per chain
           Other seams   1 ft per chain

Where the extent of the indicated coal is based on an assumed rate of thinning, the seam thickness used in making the estimate is not necessarily the full thickness anticipated from a broad knowledge of the seam and coalfield. Thus any such estimate area of indicated coal requires also an estimate of inferred coal. (NZ geol. Surv. Bull. No. 56, 1957).

indicated marginal reserve - US official procedure: "An indicated marginal reserve is estimated from the **indicated reserve base** by judging that a part of the **reserve base** has uncertainty as to economic producibility. The marginal **reserve** is computed from this uncertain part by applying an appropriate **recovery factor** based on extraction potential or by subtracting the coal that would be lost if mining should take place". (USGS 1979 Proposed Revision of Bull. 1450-B, 1976).

indicated (or probable) reserves [reservas indicadas (o probables)] - Colombian official classification: Indicated reserves are those whose quality and quantity are defined partly by means of analyses of samples and measurements and partly on the basis of reasonable geological projection.
Ingeominas considers that the observation points could be from 1 to 2.5 km apart and that indicated reserves of coal could be projected as a band up to 1 km wide beyond 0.5 km from the point of observation. The minimum information on coal quality will be of one sample per seam in each two kilometre square. (Publ. Geol. Esp. Ingeominas, No. 3, 1979).

indicated (probable) coal - Taiwan official definition: Indicated coal is coal for which the tonnage is computed partly from specific measurements and partly from projection of visible data for a reasonable distance on geologic evidence. In general the points of observation are of the order of 1000 m apart, and the outer limit of a block of indicated coal shall be of the order of 500 m from the last point of positive information. If the coal bed is believed to have greater continuity from geologic evidence, the area of measured coal is followed down the dip by a belt of indicated coal of about the same width as that of the measured coal. If any portion of a coal bed is covered by two points of observation coinciding at the same spot, e.g. drilling and mine working, the area of twice indicated coal designated by these two points of observation will become an area of measured coal. Likewise the area of twice inferred coal will become an area of indicated coal. (Bull. geol. Surv. Taiwan, No.10, 1959).

indicated reserve - Indian standard procedure: "In the case of indicated reserves, the points of observation are 1000 m (3300 ft) apart, but may be 2000 m (6600 ft) for beds of known geological continuity. Thus a line drawn 1000 to 2000 m (3300 to 6600 ft) in from an outcrop will demarcate the block of coal to be regarded as indicated". (Indian Standard Procedure for Coal Reserve Estimation, 1977).

indicated reserve - US official definition: "A category of virgin reserves having a moderate degree of geologic assurance.
Estimates of quantity are computed by projection for a specified distance beyond coal classed as **measured** of thickness, sample, and geologic data from nearby outcrops, trenches, workings, and drillholes".
- Discussion: "The thickness of coal assigned to Indicated Reserves must be considered by estimators as being currently economically extractable by means

of available technology".
- Criteria: "A Reserve in this class is estimated from the indicated reserve base by applying an appropriate recovery based on economic extraction or by subtracting the coal that is estimated will be lost during mining". (USGS 1979 Proposed Revision of Bull. 1450-B, 1976).

indicated reserve base - US official criteria: "A tonnage estimate for this category of coal consists of the sum of the estimates for indicated reserves, marginal reserves, and a part of the indicated subeconomic resources (the coal that is estimated will be lost in mining). Coal assigned to the indicated reserve base is computed by projection of thickness, depth, rank and quality data from points of measurement and sampling on the basis of geologic evidence. There are no points of measurement within coal judged as indicated. Individual points of measurement are surrounded by measured coal for 1/4 mile (0.4 km) succeeded by indicated coal from 1/4 mile (0.4 km) to 3/4 mile (1.2 km). Points for indicated coal are 1/2 mile (0.8 km) to 1 1/2 miles (2.4 km) apart. The indicated reserve base where there are many points of measurement may be projected to extend locally as a 1/2 mile (0.8 km) wide belt that lies more than 1/4 mile (0.4 km) from an outcrop or the points of measurement. This reserve base includes anthracite and bituminous coals 28 in (70 cm) or more thick and lignite and subbituminous coals 60 in (150 cm) or more thick to depths of 1000 ft (300 m)". (USGS 1979 Proposed Revision of Bull. 1450-B, 1976).

indicated reserves - NSW, Australia, definition: Are those for which the density of points of observation is sufficient to allow for a realistic estimate of the reserves, and for which there is reasonable expectation that the reserves could be raised to the measured category with further information.
1. Generally the points of observation are spaced not greater than 2 km apart.
2. Where geological conditions are favourable, indicated reserves may extend for a distance beyond workings. This distance will not be greater than 1 km from the last points of observation within the workings and the calculations should take account of known trends within the worked area. (NSW Code for Coal Reserves, 1979).

indicated reserves - Queensland, Australia, definition: Indicated coal is that for which tonnage (in tonnes) and grade are computed partly from specific measurements and samples, and partly from projection of visible data for a reasonable distance on geologic evidence. In general, the points of observation are no more than 2 km apart. Where measurements and sampling, on which computations of tonnage and grade are based, are of coal in situ or of core from a drill hole, the indicated reserves shall be designated first class. If the tonnage and grade are computed from drill hole samples other than cores, the indicated reserves shall be designated second class. Second class indicated reserves shall not be used in computing recoverable reserves. (Queensland Coal Reserves Classification, 1977).

indicated reserves - first class - : See indicated reserves -Queensland, Australia, définition.

indicated reserves - second class - : See indicated reserves -Queensland, Australia, définition.

indicated resource - US official definition: "A category of virgin demonstrated resources having a moderate degree of geologic assurance. Estimates of quantity, thickness, and extent are computed by projection for a specified distance beyond coal classed as measured of thickness, sample, and geologic data from nearby outcrops, trenches, workings and drill-holes".

- Discussion: "The thickness of coal assigned to this category must equal or exceed specified minimums as related to rank.
Estimates of the quantity of indicated resources may extend to depths of 6000 ft (1800 m) and be classed as **indicated reserves** (to 1000 ft depth) and **subeconomic resources** (to 6000 ft depth). Quality and rank may be determined from analyses of nearby samples collected from the resource body or may be inferred by projection of analytical data collected from elsewhere in the body or from adjacent bodies".
- Criteria: "A tonnage estimate for this class of coal is the sum of coal assigned to the **indicated reserve, indicated marginal reserve**, and/or part of the **subeconomic resource** categories. Such resources are computed by projection of thickness, depth, rank and quality data from points of measurement and sampling on the basis of geologic evidence. There are no points of measurement within coal judged to be indicated. Individual points of measurement are surrounded by **Measured** coal or 1/4 mile (0.4 km) succeeded by 1/2 mile (0.8 km) of indicated coal. Points of measurement for indicated coal are 1/2 (0.8 km) to 1 1/2 miles (2.4 km) apart. Indicated Resources where there are many points of measurement may be projected to extend locally as a 1/2 mile (0.8 km) wide belt that lies more than 1/4 mile (0.4 km) from an outcrop or the points of measurement. Indicated resources include **anthracite** and **bituminous** coals 14 in (35 cm) or more thick and **lignite** and **subbituminous** coals 30 in (75 cm) or more thick to depths of 6000 ft (1800 m)". (USGS 1979 Proposed Revision of Bull. 1450-B, 1976).

indicated resources - EMR, Canada, definition: These denote the precision with which given quantities of resources have been determined or estimated. They are defined as resources for which tonnages are computed partly from specific measurements and partly from reasonable geological projections. In general, the points of observation should be separated by less than the following distances Cordillera: 2000 ft (1000 ft in severely contorted areas); Plains: 1 mile; Maritimes: 2000 ft. (Canada: EMR Report EP 77 - 5, 1977).

indice de gonflement au creuset - French term: See **crucible swelling number** - UN ECE definition.

inertinite - UK standard definition: The group of **macerals** that behave in an inert or partially inert manner in carbonization processes. (BS 3323 : 1978).

inertinite - US standard definition: A group of **macerals** composed of **fusinite, semifusinite, micrinite, macrinite**, and **sclerotinite**.
Note - These macerals, even in bituminous coals, show little or no plasticity when heated and so are inert or show only a little agglutinating tendency during coking. Their reflectance is higher than that of associated **vitrinite**. (ANSI/ASTM D 2796 - 77).

inerts - ISO definition: Constituents of coal which decrease its efficiency in use, for example, **mineral matter (ash)** and **moisture** in fuel for combustion, or **fusain** in coal for carbonization. (ISO/R 1213/II: 3.01).

inerts - UK standard definition: Constituents of coal that alter its effectiveness in use. These are principally composed of mineral matter and moisture and may also contain macerals of the **inertinite** group. (BS 3323 : 1978).

inerts (Ballast) - FRG standard definition: Solid fuels contain inerts, apart from their organic material. Inerts consist of moisture and various different minerals. (DIN 51 700, 1967, para. 5.2).

inferidas, reservas - Spanish term: See **inferred (deduced, or possible) reser-

I

ves - Colombian official classification.

inferior coal - NCB, UK, definition: Inferior coal is defined as material with ash of 15.1 to 40.0%. (NCB, UK, Procedure for the Assessment of Reserves, 1972).

inferred coal - Ohai coalfield, NZ, classification: The definition of this term is as given for the Reefton Coalfield (1957, below) in the first paragraph. Thereafter there are some differences in the elaborative text and graphics. Where the extent of the **indicated coal** is based on an assumed rate of thinning, the seam thickness used in making the estimate is not necessarily the full thickness expected from a broad knowldge of the coalfield. Thus, any such estimate area of indicated coal requires, in theory, an estimate of inferred coal. In practice, because inferred estimates are based on conservative figures for the average seam thicknesses, inferred coal in an **indicated** estimate area is neglected. The limits to which inferred coal extends are based on broad geological considerations. The limits of the estimate areas are defined, and the proportion of the area in which coal of workable thickness is expected to be present is given. Where thickness of cover limits the area of coal for which estimates are made, all coal up to these limits of cover is classed as inferred. (NZ geol. Surv. Bull. No. 51, 1964).

inferred coal - Reefton coalfield, NZ, classification: This is coal for which quantitative estimates are based largely on broad knowledge of the geological character of the deposit and for which there are few, if any, samples or measurements. The estimates are based on an assumed continuity or repetition for which there is geologic evidence; this evidence may include comparison with deposits of similar type. Seams that are completely concealed may be included if there is a specific geologic evidence of their presence. Estimates of inferred coal should include a statement of the special limits within which the inferred coal may lie.
Where the extent of the **indicated coal** is based on an assumed rate of thinning the seam thickness used in making the estimate is not necessarily the full thickness anticipated from a broad knowledge of the seam and coalfield. Thus any such estimate area of **indicated coal** requires also an estimate area of inferred coal. Apart from these areas, inferred coal extends to limits based on broad geological considerations. Where thickness of cover limits the area of coal for which estimates are made, it is suggested that all coal under a cover of more than 80 per cent of the limiting thickness should be classed as inferred. (NZ geol. Surv. Bull. No. 56, 1957).

inferred (deduced or possible) reserves [reservas inferidas (deducidas, o posibles)] - Colombian official classification: This is the portion of **identified resources** the estimation of which is based principally on the known geology of the area where measurements are scarce. Ingeominas considers that the estimation of this category of reserves is based on the continuity of existence of various parameters between 2.5 and 10 km. (Publ. Geol. Esp. Ingeominas, No. 3, 1979).

inferred - US official definition, applicable to both the reserve and identified subeconomic resource components of the classification system: Coal in unexplored extensions of **demonstrated resources** for which estimates of the quality and size are based on geologic evidence and projection." - Criteria, applicable to both the reserve and subeconomic resource components of the classification system: "Quantitative estimates are based largely on broad knowledge of the geologic character of the bed or region and where few measurements of bed thickness are available. The estimates are based primarily on an assumed continuation from demonstrated coal for which there is geologic evidence. The points of observation are 1 1/2 (2.4 km) to 6 miles (9.6 km) apart. Inferred coal is projected to extend as a 2 1/4 miles (3.6 km) wide belt that lies more than 3/4 mile (1.2 km) from the outcrop or points of observation or measurement." (US geol. Surv. Bull. 1450-B, 1976).

inferred marginal reserve - US official definition: That part of an inferred reserve base, which at the time of determination borders on being producible. Estimates of quantity are based on assumed continuity from measured and indicated marginal reserves for which there is geologic evidence. Estimates are compiled by projection for specified distance beyond coal classed as an indicated marginal reserve of thickness, sample, and geologic data from distant outcrops, trenches, workings, and drill holes."
- Discussion: "The thickness of coal assigned to this category must exceed minimums as related to rank. In addition, depth must not exceed a specified maximum. Inferred marginal reserves must be considered by estimators as being marginally economically extractable by means of available technology." (USGS 1979 Proposed Revision of Bull. 1450-B, 1976).

inferred (possible) coal - Taiwan official definition: Inferred coal is coal for which quantitative estimates are based largely on broad knowledge of the geologic character of the bed, or region, and for which there are few, if any, measurements. The estimates are based on an assumed continuity for which there is geologic evidence. The coal within the established area is inferred to be present from general knowledge of the geologic character of the region. Inferred coal generally lies outside the limits defined for measured and indicated coal. The lower limit of the inferred coal varies with the dip and thickness of the coal bed and has been justified by engineering practice to be the margin of economically workable coal. Due to the complicated structural features and the great variation in coal quality and thickness at different parts of a coal bed, it is difficult to obtain accurate inferred data concerning the continuity and thickness of the coal beds. Extrapolation of information derived from the outcrops and in the present operating mines are usually found not adequately applicable to the coal beds at depth. (Bull. geol. Surv. Taiwan, No. 10, 1959).

inferred reserve - US official definition: "A category of virgin reserves having a low degree of geologic assurance. Estimates of quantity are based on assumed continuity from measured and indicated reserves for which there is geologic evidence. Estimates are computed by projection for a specified distance beyond coal classed as indicated of thickness, sample, and geologic data from distant outcrops, trenches, workings, and drill-holes".
- Discussion: "The thickness of coal assigned to this category must equal or exceed specified minimums as related to rank. In addition, the depth of burial must not exceed a specified maximum. Inferred reserves must be considered by estimators as being currently economically extractable by means of available technology".
- Criteria: "A reserve in this class is estimated from the inferred reserve base by applying an appropriate recovery factor based on economic extraction or by subtracting the coal that is estimated will be lost during mining. An exception to this definition is employed when dealing with lignite reserves". (USGS 1979 Proposed Revision of Bull. 1450-B, 1976).

inferred reserve base - US official criteria: "A tonnage estimate for this category of coal consists of the sum of the estimates for inferred reserves, inferred marginal reserves and a part of the inferred subeconomic resources. Coal assigned to the inferred reserve base is computed by projection of thickness, depth, rank and quality data from points of measurement and sampling on the basis of geologic evidence. There are no points of measurement within coal

I

judged as inferred. Individual points of measurement are surrounded by measured and indicated coal for 3/4 mile (1.2 km). Points for inferred coal are 1 1/2 miles (2.4 km) to 6 miles (9.6 km) apart. Inferred reserve base coal where there are many points of measurement may be projected to extend 2 1/4 miles (3.6 km) from an outcrop or the points of measurement. This reserve base includes wide belt that lies more than 3/4 mile (1.2 km) anthracite and bituminous coals 28 in (70 cm) or more in thickness and lignite and subbituminous coals 60 in (150 cm) or more in thickness to depths of 1000 ft (300 m)". (USGS 1979 Proposed Revision of Bull. 1450-B, 1976).

inferred reserves - Indian standard procedure: "This refers to coal for which quantitative estimates are based largely on broad knowledge of the geological character of the bed, but for which there are no measurements. The estimates are based on an assumed continuity for which there is geological evidence, and more than 1000 to 2000 m (3300 to 6600 ft) in from the outcrop". (Indian Standard Procedure for Coal Reserve Estimation, 1977).

inferred reserves - NSW, Australia, definition: Are those for which there is a poor cover of information, so that only an uncertain estimate of the reserves can be made. Further information will either raise these reserves to a higher category or show that part or all of them does not exist.
1. The term is intended for application to areas the size of a coalfield or coal province.
2. A quantitative estimate will not be allocated to inferred reserves, other than to indicate their value within the following ranges:
(a) Very large - in excess of 10 000 000 000 t.
(b) Large - 100 000 000 to 10 000 000 000 t.
(c) Small - 20 000 000 to 100 000 000 t.
(d) Very small - less than 20 000 000 t.
(NSW Code for Coal Reserves, 1979).

inferred reserves - Queensland, Australia, definition: Inferred coal is coal for which quantitative estimates are based largely on broad knowledge of the geologic character of the bed or region and for which there are few, if any, measurements or samples. The estimates are based on an assumed continuity for which there is geologic evidence. In general, inferred coal is coal lying in areas more than 1 km from points of sampling or observation used in computing reserves of higher status. A quantitative value is not allocated to inferred reserves other than to indicate its value within the following ranges.
(a) Very large - in excess of 10 000 000 000 t.
(b) Large - 100 000 000 tonnes to 10 000 000 000 t.
(c) Small - 20 000 000 tonnes to 100 000 000 t.
(d) Very small - less than 20 000 000 t.
(Queensland Coal Reserves Classification, 1977).

inferred resource - US official definition: "A category of virgin identified bodies of coal having a low degree of geologic assurance. Estimates of quantity are based on assumed continuity from measured and indicated resources for which there is geologic evidence. Estimates are computed by projection for a specified distance beyond coal classed as indicated of thickness, sample, and geologic data from distant outcrops, trenches, workings, and drill-holes".
- Discussion: "The thickness of coal assigned to this category must equal or exceed specified minimums as related to rank. In addition, the depth of burial may extend to depths of 6000 ft (1800 m) and be classed as economic, marginal economic, and subeconomic. Quality and rank are inferred from projected analytical data obtained from samples collected at distant points in the resource body or from adjacent beds".

- Criteria: "A tonnage estimate for this category is the sum of coal assigned to the inferred reserve, inferred marginal reserve and subeconomic resource categories. The estimates are quantitative and are based on projection of a broad knowledge of the geologic characters, habits, and patterns of coal beds in an area, basin, or region. Measurements and geologic data may be projected for up to 3 miles (4.8 km). Tonnage estimates are based on assumed continuation of beds beyond the indicated category. There are no points of measurement within inferred coal. Individual points of measurement are surrounded by measured and indicated coal for 3/4 mile (1.2 km) succeeded by 2 1/4 miles (3.6 km) of inferred coal. Points of measurement for inferred coal are 1 1/2 miles (2.4 km) to 6 miles (9.6 km) apart. Inferred coal where there are many points of measurement may be projected to extend locally as a 2 1/4 miles (3.6 km) wide belt that lies more than 3/4 mile (1.2 km) from an outcrop or points of measurement. Inferred resources include anthracite and bituminous coals 14 in (35 cm) or more thick and lignite and subbituminous coals 30 in (75 cm) or more thick to depths of 6000 ft (1800 m)". (USGS 1979 Proposed Revision of Bull. 1450-B, 1976).

inferred resources - EMR, Canada, definition: These denote the precision with which given quantities of resources have been determined and estimated. They are defined as resources for which quantitative estimates are based largely on broad knowledge of the geological character of the bed or region and for which few measurements of seam thickness are available. The estimates are based primarily on an assumed continuity of coal seams in areas remote from the points of observation used to calculate measured or indicated resources. (Canada: EMR Report EP 77 - 5, 1977).

inflation temperature (in Audibert-Arnu test) - Belgian standard definition: Temperature at which the piston reaches its lowest position and referred to as maximum contraction temperature. (NBN 831-05, 1970, para. 2).

inflation temperature (in Audibert - Arnu test) - UN ECE definition: Temperature at which the piston reaches its lowest point. (International Classification of Hard Coals by Type - 1956, app. III).

inherent ash - ISO definition: Ash derived from mineral matter which is not removable from coal by physical processes. (ISO/R 1213/II - 1971, 3.03).

inherent ash - UK standard definition: Ash, derived from mineral matter, that is not removable from the coal by physical means. (BS 3323 : 1978).

inherent impurity - US standard definition: The inorganic material in coal that is structurally part of the coal and cannot be separated from it by coal preparation methods. (ANSI/ASTM D 2234 - 76).

inherent moisture (in coal) - US standard definition: That moisture existing as a quality of the coal seam as it exists in its natural state of deposition and includes only that water considered to be part of the deposit and not that moisture which exists as a surface addition. To establish a finite measurement of this quality, it is essential to conform to conditions for its determination as established in Method D 1412. Inherent moisture is not to be equated with the moisture remaining after air-drying. Note - For purposes of these standards bed moisture and equilibrium moisture may be considered equivalent to inherent moisture. (ANSI/ASTM D 121 - 76).

insitu reserves - Indian standard procedure: Both in situ and recoverable reserves should be estimated in every case separately for coal in existing mines, closed mines and areas under mining leases. (Indian Standard Procedure for

I

Coal Reserve Estimation, 1977).

intercalations stériles - French term: See **dirt band** - CEC Analysis of Terms.

international steam table calorie - Indian standard definition: "Equal to 4.1868 Joules (exactly) and differs from gram calorie by approximately 0.3 part in 1000". [IS: 1350 (Part II) - 1970].

J

jhama - : Indian vernacular name for coal altered to coke by an igneous intrusion. (S.I.Tomkeieff, Coals and Bitumens, 1954).

jhama - : See burnt out coal - Indian standard procedure.

K

kontraktion - German term: See contraction - FRG standard definition.

L

laboratory sample - FRG standard definition: The laboratory sample or coarse sample, emaining after preparation of the rough sample to a size of less than 10 mm, is the sample which is packed up in an airtight way and sent directly to the laboratory for the determination of moisture, and a part of which is observed, sealed up, as arbitration sample if necessary. See **average sample** for context. (DIN 51 701, 1950, para. 2).

laboratory sample - ISO definition: The sample delivered to the laboratory for analysis or testing. (ISO/R 1213/II - 1971, 2.17).

Lagerstättenvorräte - German term: See **resources** - FRG standard procedure.

Lagerungsverhältnisse - German term: See **inclinations of strata** - CEC Analysis of Terms.

layout loss (Verlustmenge Zuschnitt; pertes au découpage) - CEC Analysis of Terms: Quantitative difference between planning reserves and layout reserves (see sketch under **working loss**). (CEC Assessment of Coal Reserves, 1980).

layout reserves (Zuschnittsvorrat; réserves au découpage) - CEC Analysis of Terms: That portion of the technically recoverable reserves which is included in a concrete seam-by-seam planning project, a minimum requirement being that planned and technically feasible workings are identifiably described in plans or maps. Forward estimates may be made on the basis of experience if no more than an indication of tonnages is required. (CEC Assessment of Coal Reserves, 1980).

lean coal - UK standard definition: Coal with a low volatile matter content, corresponding approximately to NCB code number 200. Note: This term is not used in the United Kingdom but is in general use in some other countries. (BS 3323 : 1978).

lean coal - : See **low volatile coal**.

lftr - Abbreviation for lufttrocken meaning air-dry : See **moisture in air-dried coal** - FRG standard definition.

lftr af - Abbreviation for lufttrocken und aschefrei : See **air-dry, ash-free** - FRG standard definition.

life index - CdF, France, classification: The reserves to be taken into account to define the life of a level or an entire mine are obtained by adding together a1 + p b1, where p is a coefficient of caution (always less than 1) varying from mine to mine depending on the structure of the resources. See **category "a" and "b" reserves** and **grade 1 reserves**. (CdF Reserves Classification, 1972 Model).

L

lig A - Abbreviation for lignite A : See lignite - US standard definition.

lig B - Abbreviation for lignite B : See lignite - US standard definition.

lignite - Indian standard classification:

Class and Type	Symbol	Nature	Gross CV kcal/kg (dmf)	Volatile Matter Percent (dmf)	G - K Coke Type	Moisture (60% RH) Part/100 parts Unit coal
Lignite Consolidated	L	Non-caking	6150 to 7300	> 50	A	> 20

Note: Based on available data, broad ranged of the properties are given. (IS: 770 - 1977).

lignite - ISO definition of brown coal reproduced under this heading: Until reliable parameters for differentiation of brown and hard coals are worked out and confirmed, coals considered in each country as brown, on the basis of a number of other characteristics, should be classified as brown coals regardless of their calorific value, i.e. including the cases when the gross calorific value of the coal in equilibrium with air at 30°C and 96% relative humidity is more than 24000 kJ/kg on the ash-free basis.

Group parameter; tar yield on the dry, ash-free basis %	Group number	Code numbers					
25	4	14	24	34	44	54	64
20 to 25	3	13	23	33	43	53	63
15 to 20	2	12	22	32	42	52	62
10 to 15	1	11	21	31	41	51	61
10 or less	0	10	20	30	40	50	60
Class number		1	2	3	4	5	6
Class parameter: total moisture content of run-of-mine coal on the ash-free basis, %.		20 or less	20 to 30	30 to 40	40 to 50	50 to 60	60 to 70

(ISO 2950, 1974).

lignite - UK standard definition: Coal of low rank with high inherent moisture content and high volatile matter content. The general term lignite may be applied to black lignite, brown lignite and **brown coal**. Note: These coals are outside the range of the NCB classification system and are not commercially mined in the United Kingdom. (BS 3323 : 1978).

lignite - US standard definition:

Class	Group	Calorific Value Limits Btu per pound (Moist, * Mineral-Matter-Free Basis)		Agglomerating Character
		Equal to or more than	Less than	
IV. Lignitic	1. Lignite A 2. Lignite B	6300 ...	8300 6300	Non-agglomerating

* Moist refers to coal containing its natural inherent moisture but not including visible water on the surface of the coal. (ANSI/ASTM D 388 - 77).

lignites (ligniti) - Italian classification: Having a **gross calorific value** less than 5700 kcal/kg on the basis of **mineral-matter-free** coal in equilibrium with air at 30°C and 96% relative humidity, they are subdivided into **peat, xyloid, and pitch lignites** (ligniti torbose, xiloidi and picee on the basis of the run-of-mine **lower (or net) calorific value**. The latter parameter is of greater interest since the main use of lignites is in power stations generally located at the mine. In addition, the quantity of constitutional ash diminishes in the passage from peat to pitch lignites, a fact which contributes to an increase in the calorific value of the run-of-mine product. (ENI-ENEL-FINSIDER Classification Systems for Coals, 1978).

ligniti - Italian term: See **lignites** - Italian classification.

ligniti picee - Italian term: See **pitch lignites** - Italian classification.

ligniti torbose - Italian term: See **peat lignites** - Italian classification.

ligniti xiloidi - Italian term: See **xyloid lignites** - Italian classification.

litantraci - Italian term: See **bituminous coal** - Italian classification.

litantraci alto volatili A - Italian term: See **high volatile A bituminous coal** - Italian classification:

litantraci alto volatili B - Italian term: See **high volatile B bituminous coal** - Italian classification.

litantraci basso volatili - Italian term: See **low volatile bituminous coals** Italian classification.

litantraci medio volatili - Italian term: See **medium volatile bituminous coals** - Italian classification.

lithotype - US standard definition: Any of the banded constituents of coal: vitrain, fusain, clarain, durain or attrital coal, or a specific mixture of two or more of these. (ANSI/ASTM D 2796 - 77).

L

long flame coal - UK standard definition: Coal that has a high volatile matter content and is generally free-burning. Note: This term is used in many European countries. (BS 3323 : 1978).

loss factor (Verlustmengenfaktor; facteur de pertes) - CEC Analysis of Terms: Dimensionless factor applied to reserves in order to indicate the proportion of losses. This factor is always o < f < 1. (CEC Assessment of Coal Reserves, 1980).

low volatile bituminous - : See entries for bituminous coal and low volatile coal.

low volatile bituminous coal - US standard classification.

low volatile bituminous coal (litantraci basso volatili - Italian Classification. These coals have a volatile matter content from 10% to 20% on a dry, ash-free basis, within the general classification of hard coals. Their index of reflectivity of vitrinite is greater than 1.5%. They have a high coking power when their dilatation is positive and greater than 20% and their free swelling index from 7 to 9 medium coking power when their dilatation is positive from 0 to 20% and free swelling index from 4 to 6.5; low coking power when their dilatation is negative and free swelling index from 1.5 to 3.5 and no coking power when they only contract in the dilatometer test and have a free swelling index from 0 to 1. (ENI-ENEL-FINSIDER Classification Systems for Coals, 1978).

low volatile coal - NCB, UK, classification:

General Description	Coal Rank Code			Volatile Matter (dmmf) (per cent.)	Gray-King Coke Type *
	Main Class	Class	Sub-Class		
Low-volatile steam coals	200			9.1-19.5	A-G8
Dry steam coals		201	201 a 201 b	9.1-13.5 9.1-11.5 11.6-13.5	A-C A-B B-C
Coking steam coals		202 203 204		13.6-15.0 15.1-17.0 17.1-19.5	B-G E-G4 G1-G8

* Coals with **volatile matter** of under 19.6 per cent. are classified by using the parameter of volatile matter alone; the Gray-King coke types quoted for these coals indicate the general ranges found in practice, and are not criteria for classification. (NCB Coal Classification System, 1964).

low volatile coal - RAG, FRG definition: Magerkohle (lean coal): V.M. (daf) by weight: 10% to 14%; Crucible coke: Powdery. Esskohle (steam coal): V.M. (daf) by weight: 14% to 20%; Crucible coke lightly caked to sintered. (Ruhrkohlen-Hanbdbuch, Para. 1.1.1.1., Table, 1969).

low volatile coal - UK standard definition: Coal with a volatile matter content (DMMF) of up to 19.5%. Note: It corresponds to NCB rank codes 100 and 200. (BS 3323 : 1978).

L

low volatile coal - : See also bituminous coals.

low volatile coals (Magerkohle & geringbituminöse Kohle) - FRG standard definition:

Name (indication)	International Class (first digit in index no.)	Volatile Matter content in dry ash-free matter % by weight
Magerkohle Lean coal	2	> 10 to 14
geringbituminöse Kohle Low volatile bituminous coal	3 a b	> 14 to 16 > 16 to 20

(DIN 23 003, 1976, Table 1).

lufttrocken (lftr) - German term meaning **air-dry**: See **moisture in air-dried coal** - FRG standard definition.

lufttrocken und aschefrei (lftr af) - German term: See **air-dry, ash-free** - FRG standard definition.

lvb - Abbreviation for **low volatile bituminous** : See **low volatile bituminous coal** - US standard definition.

M

ma - Abbreviation for **meta-anthracite**: See **anthracite** - US standard definition.

maceral - US standard definition: A microscopically distinguishable organic component of coal, but including any mineral matter not discernable under the optical microscope. Note - Macerals are recognized on the basis of their reflectance and morphology. A given maceral may differ significantly in composition and other properties from one coal to another; for some macerals the variation depends mostly on the rank of the coal. Inorganic impurities of submicroscopic in size, considered to be part of the maceral, may amount to several percent attrital coal. (ANSI/ASTM D 2796 - 77).

maceral group - US standard definition: The classification of the microscopic constituents into groups of similar properties in a given coal is as follows:

Maceral Group	Maceral
Vitrinite	vitrinite
Exinite (or Liptinite)	alginite
	cutinite
	resinite
	sporinite
Inertinite	fusinite
	macrinite
	sclerotinite
	semifusinite

(ANSI/ASTM D 2796 - 77, para. 5.1).

macerals - UK standard definition: The microscopically recognizable individual organic constituents of coal. Note: There are three groups of macerals, namely vitrinite, exinite and inertinite. (BS 3323 : 1978).

Mächtigkeit - German term: See **thickness** - CEC Analysis of Terms.

macrinite - US standard definition: The maceral that is distinguished by a reflectance higher than that of associated vitrinite, absence of recognizable plant cell structure, and by a particle size 10 µm or greater on any diameter of the section under examination. (ANSI/ASTM D 2796 - 77).

Magerkohle - German term meaning **lean coal** : See **low volatile coal** - FRG standard definition.

marginal reserve - US official definition: "That part of a **reserve base**, which at the time of determination, borders on being producible. A marginal reserve has the essential characteristic of uncertainty as to economic producibility. Included in a marginal reserve is that part of a **resource** that would be economically producible given projected changes in economic or technologic factors and that part that is left in the ground during extraction of a reserve. If

- 70 -

the part left in the ground is recoverable in a second phase of extraction, it would be classed as a marginal reserve." (USGS 1979 Proposed Revision of Bull. 1450-B, 1976).

marketable reserves - NSW, Australia, definition: Consist of the amount of coal available for sale after marketing factors, e.g. coal treatment, have been considered. (NSW Code for Coal Reserves, 1979).

marketable reserves (riserve commerciali) - Italian classification: "These are the quantities of coal effectively available for marketing". (ENI-ENEL-FIN-SIDER Classification Systems for Coal, 1978).

matieres volatiles - French term (abbreviation M.V.): See volatile matter - Belgian standard definition.

maximum contraction temperature - (in Audibert-Arnu test): See inflation temperature - Belgian standard definition.

maximum dilatation temperature - (in Audibert-Arnu test): See temperature of maximum dilatation.

mean gram calorie - Indian standard definition: "One hundredth part of the amount of heat required to raise the temperature of one gram of water from 0° to 100°C. It is equal to 4.1897 joules". [IS: 1350 (Part II) - 1970].

measured - US official definition, applicable to both the reserve and identified subeconomic resource components of the classification system: "Coal for which estimates of the rank, quality, and quantity have been computed, within a margin of error of less than 20 percent, from sample analyses and measurements from closely spaced and geologically well-known sample sites".
- Criteria, applicable to both the reserve and subeconomic resource components of the classification system: Resources are computed from dimensions revealed in outcrops, trenches, mine workings, and drill holes. The points of observation and measurement are so closely spaced and the thickness and extent of coals are so well defined that the tonnage is judged to be accurate within 20 percent of true tonnage. Although the spacing of the points of observation necessary to demonstrate continuity of the coal differs from region to region according to the character of the coal beds, the points of observation are no greater than 1/2 mile (0.8 km) apart. Measured coal is projected to extend as a 1/4 mile (0.4 km) wide belt from the outcrop or points of observation or measurement". (US geol. Surv. Bull. 1450-B, 1976).

measured coal - Ohai coalfield, NZ, definition: The definition/explanation of this term is as given for the Reefton Coalfield (1957, below), but with the addition of the following text and with some amendment to the graphical elaboration, as follows: The ash contents of samples from drill holes are not regarded as being representative of the ash contents of the seams from which they are derived, unless the percentage of core recovery is high. (NZ geol. Surv. Bull. No. 51, 1964).

measured coal - Reefton coalfield, NZ, definition: This is coal for which tonnage is computed from dimensions revealed in outcrops, trenches, workings, and drill holes and for which the ash content is computed from the results of detailed sampling. The sites for inspection, sampling, and measurement are so closely spaced and the geologic character is defined so well that the size and shape are well established. The computed tonnage and ash content are judged to be accurate within limits which are stated, and no such limit is judged to differ from the computed tonnage or ash content by more than 20 per cent. (NZ

M

geol. Surv. Bull. No. 56, 1957).

measured (proved) coal - Taiwan official definition: Measured coal is coal for which tonnage is computed from dimensions revealed in outcrops, trenches, mine workings, and drill holes. The spacing of these points of observation is generally in the order of 500 m apart. The outer limit of a block of measured coal shall be in the order of 250 m from the last point of positive information. If any portion of a coal bed is covered by two points of observation coinciding at the same spot, e.g. drilling and mine working, the area of twice the **indicated coal** designated by these two points of observation will become an area of measured coal. (Bull. geol. Surv. Taiwan, No. 10, 1959).

measured reserve - US official definition: "A category of accessed and virgin **reserves** having a high degree of geologic assurance. Estimates of quantity are computed, partly from measurements and sample analyses obtained at geologically well-defined points of measurement and sampling, and partly by projection of data not exceeding a specified distance".
- Discussion: "Measured reserves have the highest degree of geologic assurance and abut the points of measurement and sampling. Samples obtained at outcrops, in trenches, and from mine-workings and drill-holes. The thickness of coal assigned to this class must equal or exceed specified minimums as related to rank. In addition, the depth of burial must not exceed a specified maximum for each rank. Measured reserves must be considered by estimators as being currently economically extractable by means of available technology."
- Criteria: "A reserve in this class is estimated from the **measured reserve base** by applying an appropriate recovery factor based on economic extraction or by subtracting the coal that it is estimated will be lost during mining (See criteria for various classes of Reserves)". (USGS 1979 Proposed Revision of Bull. 1450-B, 1976).

measured reserve base - US official criteria: "A tonnage estimate for this class of coal consists of the sum of the estimates for **measured reserves, marginal reserves**, and a part of the **measured subeconomic resources** (the coal that is estimated will be lost in mining). Coal assigned to the measured reserve base is computed by projection of thickness, depth, rank, and quality data from points of measurement and sampling on the basis of geologic evidence for not more than 1/4 mile (0.4 km). Thus, individual points of measurement are up to 1/2 mile (0.8 km) apart. Where there are many nearby points of measurement coal assigned to a measured reserve may be projected locally as a 1/4 mile (0.4 km) wide belt that lies adjacent to the points of measurement and an outcrop. The reserve base includes **anthracite** and **bituminous** coals 28 in (70 cm) or more thick and **subbituminous coals** 60 in (150 cm) or more thick to depths of 1000 ft (300 m) and **lignite** beds 60 in (150 cm) or more thick to depths of 300 ft (100 m)". (USGS 1979 Proposed Revision of Bull. 1450-B, 1976).

measured (or proved) reserves [reservas medidas (o probadas)] - Colombian official classification: Measured reserves are those whose quality and quantity have been estimated within a minimum margin of error of 20% by means of closely spaced analyses and measurements from outcrops, trenches, drill holes or workings so that the thickness and extent of the mineralised body are defined with the best possible accuracy. INGEOMINAS, in the conduct of its own exploration considers that, to calculate measured coal reserves, the points of observation and control should be not less than 1 km apart. Also the measurements between control points may be extrapolated up to 0.5 km from the last points of observation. The minimum information on quality will be of one sample per kilometre square from each coal seam, restricted by INGEOMINAS to economically exploitable beds. (Publ. Geol. Esp. Ingeominas, No. 3, 1979).

measured reserves - NSW, Australia, definition:- are those for which the density of points of observation is sufficient to give control on quality, thickness, depth, and other relevant conditions, and to allow for both a reliable estimate of the reserves and the planning of their extraction.
1. Experience shows that the above conditions can rarely be satisfied if the points of observation are spaced at a distance greater than 1 km. Often a greater density will be required.
2. Where geological conditions are favourable it may be possible to assume knowledge of reserves for a distance from workings without additional observation points by extrapolation of known trends within the worked area. In such cases the distance will be not greater than 0.5 km from the last points of observation within the workings. (NSW Code for Coal Reserves, 1979).

measured reserves - Queensland, Australia definition: Measured coal is that for which tonnage (in tonnes) is computed from dimensions revealted in outcrops, trenches, mine workings, and drill holes and for which the grade is computed from the results of detailed sampling. The points of observation, measurement, and sampling are so closely spaced and the thickness, extent, and grade of the coal are so well defined that the computed tonnage and grade are judged to be accurate within 20 per cent. of the true tonnage and grade. The limits of accuracy of the estimate or the maximum distance between points of observation should be stated. Although the spacing of the points of observation necessary to demonstrate continuity of coal will vary in different regions according to the habit of the coal seams, the points of observation are no more than 1 km apart. The outer limit of a block of measured coal, therefore, shall not be more than 0.5 km from the last point of positive information (i.e., about half the distance between points of observation). The Standards Association of Australia considers that, where a point of observation is a drill hole, core only is to be regarded as a true sample. Consequently only core shall be used as a drill hole sample in computing measured reserves. Minimum core recovery at any point of observation, which should exceed 90 per cent., should be stated. (Queensland Coal Reserves Classification, 1977).

measured resource - US official definition: "A category of accessed and virgin demonstrated resources having a high degree of geologic assurance. Estimates of quantity, thickness, and extent are computed partly from the analysis of measurements and samples obtained at geologically well-defined points of measurement and sampling and partly by projection of data not exceeding a specified distance."
- Discussion: "Points of measurement and sampling are not to be more than a specified distance apart. Estimates of quantity of measured resources may extend to depths of 6000 ft (1800 m) and be classed as **measured reserves** (to 1000 ft depth), **marginal reserves** (to 1000 ft depth), and **subeconomic resources** (to 6000 ft depth). Quality and rank may be determined from analyses of samples collected from the measured coal or inferred by projection of analytical data collected elsewhere from the same resource body or from adjacent beds".
- Criteria: "A tonnage estimate for this class of coal is the sum of the coal assigned to the **measured reserve, measured marginal reserve,** and a part of the **subeconomic measured resources** categories. Such resources are computed from measurements and by projection of thickness, depth, rank, and quality data not to exceed 1/4 mile (0.4 km). Points of measurement, thus, are up to 1/2 mile (0.8 km) apart. Measured resources where there are many points of measurement may be projected to extend locally as a 1/4 mile (0.4 km) wide belt from an outcrop or the points of measurement. Measured resources include **anthracites** and **bituminous coals** of 14 in (35 cm) or more thick and **lignite** and **subbituminous coals** 30 in (75 cm) or more thick and to depths of 6000 ft (1800 m)". (USGS 1979 Proposed Revision of Bull. 1450-B, 1976).

M

measured resources - EMR, Canada, definition:- denote the precision with which given quantities of resources have been determined or estimated. They are defined as **resources** for which tonnages are computed from dimensions revealed in outcrops, trenches, mine workings and boreholes. The spacing of points of observation necessary to justify confidence in the continuity of coal seams differs from region to region according to the character of the seams. In general, the points of observation should be separated by less than the following distances:
(i)   Cordillera, 1000 ft (500 ft in severely contorted areas);
(ii) Plains, 1/2 mile;
(iii) Maritimes, 1,000 ft.
(EMR Report EP 77 - 5, 1977).

medidas, reservas - Spanish term: See **measured (or proved) reserves** - Colombian standard classification.

medium volatile bituminous - : See entries for **bituminous coal** and **medium volatile coal**.

medium volatile bituminous coals (litantraci medio volatili) - Italian classification. These coal have a volatile matter content from 20% to 28% on a dry, ash-free basis within the general classification of hard coals and an index of reflectivity of **vitrinite** from 1.1% to 1.5%. They have a high coking power when their dilatation is positive and more than 80% with a free swelling index from 7 to 9; medium coking power when their dilatation is positive from 30% to 80% with a free swelling index from 4 to 6.5; low coking power when their dilatation ranges from negative up to a positive dilatation of 30%, with a free swelling index from 1.5 to 3.5; and no coking power when they only contract in the dilatometer test and have a free swelling index from 0 to 1. (ENI-ENEL-FINSIDER Classification Systems for Coals, 1979).

medium volatile bituminous coal (litantraci medio volatili) - Italian classification.

medium volatile bituminous (mvb) - See **medium volatile coals** - US standard definition.

medium volatile coal (mittelbituminöse Kohle) : - FRG standard definition:

Name (indication)	International class (first digit in index no.)	Volatile Matter Content in dry ash-free matter % by weight
mittelbituminöse Kohle Medium volatile bituminous coal	4	> 20 to 28

(DIN 23 003, 1976, Table 1).

medium volatile coal - : See also **bituminous** and **high volatile coal**.

- 74 -

M

medium volatile coal - NCB, UK, classification:

Coal Rank Code			Volatile Matter (d.m.m.f.) (per cent.)	Gray-King Coke Type	General Description
Main Class	Class	Sub-class			
300			19.6-32.0	A-G9 and over	Medium-volatile coals
	301		19.6-32.0	G4 and over	
		301 a	19.6-27.5	)G4 and	Prime coking coals
		301 b	27.6-32.0	)over	
	302		19.6-32.0	G-G3	Medium-volatile, medium-caking or weakly caking coals
	303		19.6-32.0	A-F	Medium-volatile, weakly caking to non-caking coals

(NCB Coal Classification System, 1964).

medium volatile coal - Polish standard classification:

Coal Type		Classification criteria			
Name	Code Number	Volatile Matter (daf) %	Caking Properties Roga Index	Dilatation	General technical characteristics
Coking Coal	35.1	Over 26 up to 31	Over 45	Over 30	Medium volatile matter content; high caking properties, positive dilatation
Wegiel ortokoksowy	35.2	Over 20 up to 26		Over 0	
Coking coal; Wegiel metakoksowy	36	Over 14 up to 20	Over 45	Over 0	Medium volatile matter content; good caking properties, positive dilatation
Semi Coking coal; Wegiel semikoksowy	37	Over 14 up to 28	5 or over	Not in standard	Medium volatile matter content; poor caking properties
Lean coal Wegiel chudy	38	Over 14 up to 28	Below 5	Not in standard	Medium volatile matter content; virtually no caking properties

Heat of combustion not among classification criteria for medium volatile coal in this standard.
Volatile matter determined according to Polish standard PN-66/G-04516 (Hard coal);

M

　　Caking properties determined according to Polish standard PN-67/G-04518 (Hard coal);
　　Dilatation determined according to Polish standard PN-66/G-04517 (Hard coal); (PN-68/G-97002).

medium volatile coal - UK standard definition: Coal with a volatile matter content (DMMF) from 19.6% to 32% Note: It corresponds to NCB rank code 300. (BS 3323 : 1978).

medium volatile coal - US standard definition:

Class	Group	Fixed Carbon Limits. percent (Dry, Mineral-Matter-Free-Basis)		Volatile Matter Limits. percent (Dry, Mineral-Matter-Free-Basis)		Agglomerating Character
		Equal or Greater Than	Less Than	Greater Than	Equal or Less Than	
1. Bituminous	2. Medium volatile bituminous coal	69	78	22	31	commonly agglomerating

(ANSI/ASTM D 388 - 77).

Menge - German term: See **tonnage** - CEC Analysis of Terms.

meta-anthracite (ma) - : See **anthracite** - US standard definition.

Meta-Anthrazit - German term for **meta-anthracite**: See anthracite - FRG standard definition.

metakoksowy wegiel - Polish term for a class of coking coal: See **medium volatile coal** - Polish standard classification.

micrinite - US standard definition: The maceral that is distinguished by reflectance higher than that of associated **vitrinite**, absence of recognizable plant cell structure, and occurring as dispersed or aggregated particles of size less than 10 µm and commonly about 1 µm (ANSI/ASTM D 2796 - 77).

mine verification line - : See **reserves calculation method** - Japanese standard procedure.

mineable coal (reserve level 1) - EMR, Canada, classification: This is derived from measured and indicated **resources** by qualification in terms of current technological limits and value of the final product. It is an in-place tonnage which could represent less than 20% of the initial resource.
(Canada: EMR Report EP 77 - 5, 1977).

mineable coal in situ - Rep. South Africa, definition: that portion of the coal in situ which can be mined by existing techniques. Table 5.31 of ref. specifies condition of coal in situ in terms of depth and workable thickness range; minimum yield in preparation plant, ash, volatile matter and swelling indices; by types of coal. See also - coal in situ, extractable coal. (Dept. Mines, Rept.

Coal Res. SA 1975).

mineable resources (bergbauliche Vorräte) - FRG standard definition: For brown coal, this category includes all resources which are assumed to fulfil present requirements for economic exploitation. These requirements relate to qualities, form, thickness and volume. (DIN 21 942, 1961).

mineral matter - UK standard definition: The inorganic material in coal. Note: It includes **water of constitution** but excludes **total moisture** and **moisture in the analysis sample**. (BS 3323 : 1978).

mineral-matter-free - : See **dry mineral-matter-free (dmmf)**; **moist mineral-matter-free (mmmf)**.

mineral parting - US standard definition: Discrete layer of mineral or mineral-dominated sediment interbedded with coal along which, in mining, separation commonly occurs. Layers of **bone** coal having indefinite boundaries usually are not considered to be partings because they do not form planes of physical weakness. They may merge vertically or horizontally with layers that are bony or coaly shale and that do form planes of physical weakness.
(ANSI/ASTM D 2796 - 77).

mineralized coal - US standard definition: Impure coal that is heavily impregnated with mineral matter, either dispersed or discretely localized along cleat joints or other fissures. Pyritic or calcareous mineralized coal is most common. (ANSI/ASTM D 2796 - 77).

Mineralstoff - German term for **mineral matter** : See, for example, **dry, mineral-matter-free** - FRG standard definition.

mining losses - Indian standard procedure: For coal in existing mines, closed mines and areas under mining leases, the possible losses in working shall be given under the following heads:-
(a) Coal likely to be lost due to geological features;
(b) Coal likely to be locked up under roads, railway lines and rivers;
(c) Coals likely to be lost in barriers;
(d) Mining losses.
The working plans of the collieries and the latest geological maps of the coalfields shall form the basis of the calculations. (Indian Standard Procedure for Coal Reserve Estimation, 1977).

mining losses (Abbauverluste) - FRG standard procedure: For brown coal, mining losses are dependent on deposit conditions and the particular mining method employed.
  i) The following are to be considered when calculating surface **mining resources**.
     Stripping losses (Schnittverluste), incurred at each interface between dirt and coal, and losses which arise from the separation of partial impurities in the deposit. Stripping losses can be extremely significant for the economic exploitability of an occurrence with severe seam splitting or a concentrated sequence of numerous thin seams.
  ii) In calculations of **deep mineable resources**, mining losses consist of losses at roof and floor, of wedge losses when individual cuts reach the roof or floor in dipping deposits, of losses between each cut and of pillar losses within each cut.
The average thickness of the stripping losses at the interface between coal and dirt is given in metres. The losses probably to be expected as a result of deposit disturbance are given in %. If there are no empirical values available,

M

estimated values must be assumed. If further mining losses are to be expected for other reasons, these must be reported with their cause. The total mining losses from the sum of all separate aspects is reported as a percentage. (DIN 21 942, 1961).

mining losses (perdite di coltivazione) - Italian definition: "These are those parts of in situ reserves which cannot be exploited because of intrinsic characteristics of the mining method and because of environmental restraints". (ENI-ENEL-FINSIDER Classification Systems for Coals, 1978).

mining recovery factors - Indian standard procedure: Coal mined and lost in mining shall be calculated by either of the two methods given below:
i) By actual quantitative measurements in working mines. Its comparison with coal production figues gives a yield percentage of recovery data which can be applied under similar conditions in virgin areas to determine the recoverable reserves.
ii) Where no data is available, the production figures, increased by a percentage factor of losses in mining, gives the quantity of coal worked out in a particular coalfield. The balance are the available reserves which will give the recoverable reserves where no other data is available the yield percentage factor may be taken as 50 tonnes for every 100 tonnes of coal mined. (Indian Standard Procedure for Coal Reserve Estimation, 1977).

mining resources (abbaufähige Vorräte) - FRG standard procedure: For brown coal, coal, the mining resources are calculated from the proved, probable and possible class after subtracting all resources in berms, batters (or walls) and safety pillars which cannot be mined because of surface safety or other reasons. (DIN 21 942, 1961).

mittlere Mächtigkeit - German term: See average thickness - CEC Analysis of Terms.

mmmf - Abbreviation: See moist, mineral-matter-free.

mogelijke - Flemish term: See possible reserves - Belgian classification.

mögliche Vorräte - German term: See possible resources - FRG standard definition.

moist, mineral-matter-free - US standard procedure: For classification of coal according to rank,.... calorific value shall be calculated to the mineral-matter-free basis in accordance with either the Parr.... or the approximation formulas, ... that follow. In case of litigation use the appropriate Parr formula. Parr Formula:
Moist, MM-free Btu = (Btu - 50S) / [100 - (1.08A + 0.55S)] x 100.
Approximation Formula: Moist, MM-free Btu = Btu / [100 - (1.1A + 0.1S)] x 100
Where:
MM = mineral matter
Btu = British thermal units per pound (calorific value)
A = percentage of ash, and
S = percentage of sulphur.
Above quantities are all on the inherent moisture basis. This basis refers to coal containing its natural inherent or bed moisture but not including water adhering to the surface of the coal. See Method D 3180, 5.1.3, for calculating analyses to this basis. (ANSI/ASTM D 388 - 77, para. 8).

moisture - Indian standard definition: "Water expelled in its various forms when tested under the specified conditions". [IS: 1350 (Part I) 1969].

moisture - : See raw (roh); analysis moist (analysenfeucht); dry (wasserfrei); dry, ash-free (wasser- und aschefrei); dry, mineral-matter-free (wasser- und mineralstofffrei) - FRG standard definition. The distinctions between these terms involving moisture are shown in the schematic representation of the relative analytic states of solid fuels. (DIN 51 700, 1967, para. 5.2).

moisture-free (wasserfrei) - : See dry - FRG standard definition.

moisture-holding capacity - UK standard definition: The quantity of moisture contained by a coal at a temperature of 30°C in equilibrium with an atmosphere saturated with water vapour (96% rh) (BS 3323 : 1978).

moisture, in air-dried coal - Australian standard definition: That part of the total moisture which is retained by the coal after it has been exposed to the atmosphere and has gained approximate equilibrium. (AS K184 - 1969, para. 3.4).

moisture in air-dried coal - UK standard definition: The moisture in the coal sample after it has attained equilibrium with ambient laboratory conditions. (BS 3323 : 1978).

moisture in air-dried coal; air-dry (lufttrocken; lftr) - FRG standard definition: Fuel is known as "a-dry" * if there is no surface moisture present. It thus appears dry. In this condition the moisture content of the fuel has adjusted itself to atmospheric humidity and is in equilibrium with this. The air-dry condition is reached by storing fuel in the air. Wet fuels lose moisture through evaporation, dry fuels take up moisture from the air. The degree of saturation depends on relative atmospheric humidty and on temperature. Thus the moisture content of air-dry fuel stays constant if the storage conditions remain unchanged. If the degree of saturation is to serve as the characteristic indicator of the capacity to absorb water, equilibrium conditions must be established. 30°C and a relative atmospheric humidity of 97% have been agreed as equilibrium conditions for high volatile coal classes in the International Classification System for Hard Coals.
  * To date the "analysis moist" fuel has been called "airdry" in the sphere of fuel analysis. Since this designation does not correspond to the meaning of the term "air-dry", the condition "analysis-moist" has been introduced. In place of the indices (lftr) or (i.lftr) as used in previous instructions the indices (an) or (i.an) are used in new instructions. (DIN 51 700, 1967, para. 5.2).

moisture in air-dried coal (air drying; air dry loss) - US standard definitions:
1) Air drying - a process of partial drying of coal to bring its moisture near to equilibrium with the atmosphere in the room in which further reduction and division of the sample is to take place. (ANSI/ASTM D 2013 - 72, para. 4.1).
2) Air dry loss - the loss in weight, expressed in percentage, resulting from the partial drying of coal at each stage of reduction or division. (ASTM D 3302, - 74, para. 5.2).

moisture in analysis sample - FRG standard definitions:
1) The moisture content of the ready-ground fuel as it is being weighed for analysis is known as analysis moisture. In this condition the fuel is analysis moist (analysenfeucht). In the analysis moist condition there is no surface moisture on the fuel, and it can be sieved after grinding. For practical analytical purposes fuels can generally be completely dried out before investigation. For scientific investigations of fuels which are sensitive to oxygen and high temperatures the water may, however, only be evaporated off if changes in the technological behaviour and in composition are to be avoided. The remainder,

after drying, is analysis moist fuel, the moisture content of which is more or less equivalent to atmospheric humidity. (DIN 51 700, 1967, para. 5.2).
2) Analysis moisture is the **hygroscopic moisture** possessed by an air-dry sample for analysis according to DIN 51 700 dependent on the storage conditions at the time of each of the analytical investigations. (DIN 51 718, 1978).

moisture in analysis sample - ISO definition: The moisture in the analysis in ISO Recommendation R 331 'Determination of moisture in the analysis sample of coal by the direct gravimetric method' and ISO Recommendation R 348 'Determination of moisture in the analysis sample of coal by the direct volumetric method'. (ISO/R 1213/II - 1971, 3.31).

moisture in the analysis sample - UK standard definition: The moisture in the **analysis sample** of coal (ground to pass a 212 μm test sieve) after it has attained equilibrium with ambient laboratory conditions. (BS 3323 : 1978).

moisture sample - FRG standard definition: This is a sample which is taken away from the **rough sample** - the latter having been crushed to a size of less than 20 mm - in cases where a knowledge of the original water content is specially wanted and a loss of water is to be expected at further crushing. See **average sample** for context. (DIN 51 701, 1950, para. 2).

moisture sample - ISO definition: A sample used exclusively for the purpose of determining moisture. (ISO/R 1213/II - 1971, 2.19).

M.V. - French abbreviation for matieres volatiles: See **volatile matter** - Belgian standard definition.

mvb - Abbreviation for **medium volatile bituminous**: See **medium volatile coal** - US standard definition.

N

net calorific value (Heizwert) - FRG standard definition: The net calorific value H(u) is the ratio of the amount of heat generated by the complete combustion of a solid or liquid fuel to the weight of the fuel, when:
a) the temperature of the fuel before combustion and that of products of combustion is 25°C;
b) the moisture present in the fuel prior to combustion and the moisture formed by the combustion of hydrogen fuels remains as vapour;
c) the products of combustion, carbon dioxide, and sulphur dioxide remain in a gaseous state; and
d) the nitrogen has not oxidised. (DIN 51 900, Part 1, 1977, para. 3.3).

net calorific value - Indian standard definition: "Number of heat units liberated when a unit mass of the fuel is burnt at constant volume in oxygen saturated with water vapour, the original and final materials being at approximately 25°C. The residual products are taken as carbon dioxide, sulphur dioxide, nittrogen and water vapour". [IS: 1350 (Part II) - 1970].

net calorific value - ISO definition: The **gross calorific value** less the latent heat of evaporation of the water originally contained in the fuel and that formed during its combustion. (ISO/R 1213/II 3.13).

net calorific value at constant pressure - UK standard definition: The number of heat units that would be liberated if unit mass of fuel was burned in oxygen at constant pressure in such a way that the heat release was equal to the gross calorific value of the fuel at constant pressure, less the latent heat of evaporation of the water, at 25°C at constant pressure. Note: The deduction of the latent heat evaporation of water applies to both the moisture in the fuel and that formed by combustion. (BS 3323 : 1978).

net calorific value, H(i) - US standard definition: In case of solid fuels and liquid fuels of low volatility, a lower value calculated from the **gross calorific value** as the heat produced by combustion of unit quantity, at constant atmospheric pressure, under conditions such that all water in the products remain in the form of vapour. Note: The net calorific value (net heat of combustion) is calculated from the gross calorific value [gross heat of combustion at 68°F (20°C) by making a deduction of 1030 Btu/lb (572 cal/g)] of water derived from unit quantity of fuel, including both the water originally present as moisture and that formed by combustion. The deduction is not equal to the latent heat of vaporisation of water (1055 Btu/lb at 68°F (20°C)) because the calculation is made to reduce from gross value at constant volume to net value at constant pressure, for which the appropriate factor under these conditions is 1030 Btu/lb. (ANSI/ASTM D 121 - 76).

net heat of combustion - : See **net calorific value** - US standard definition.

net (specific) calorific value [Heizwert (spezifischer); pouvoir calorifique

N

(spécifique)] - CEC Analysis of Terms: The gross (specific) calorific value less the heat vaporisation (vaporisation enthalpy in the case of combustion under constant pressure) of the water contained in the fuel (at 25°C). The specific enthalpy of vaporisation of water at 25°C is 2442 kJ/kg = 583 kcal/kg. Standard: Internationally ISO/R 1928-1971. (CEC Assessment of Coal Reserves, 1980).

net thickness (Nettomächtigkeit; puissance nette) - CEC Analysis of Terms: Total thickness of recoverable strata excluding intermediate rock and dirt bands in a deposit or part of a deposit. (CEC Assessment of Coal Reserves, 1980).

Nettomächtigkeit - German term: See **net thickness** - CEC Analysis of Terms.

nicht abbauwürdig - German term: See **uneconomic** - Austrian guidelines.

nonbanded coal - US standard definition: Consistently fine-granular coal essentially devoid of megascopic bands. (ANSI/ASTM D 2796 - 77).

not presently mineable reserves (riserve non coltivabili attualmente) - Italian classification: "These are all **proved**, **probable** and **possible reserves** having characteristics which do not permit economic mining at the time of determination. (ENI-ENEL-FINSIDER Classification Systems for Coals, 1978).

O

oblique distance method - : See reserves calculation method - Japanese standard procedure.

occurrences - UN CNR classification: Any additional material with a lower economic potential outside the boundaries of resources should be referred to as occurrences and should be reported separately. (UN CNR International Classification of Mineral Resources, 1979).

operating thickness overburden ratio (A:K, Abraum:Kohle) - FRG standard procedure: For brown coal, is obtained by substituting the "in place" thickness by the "useful" thickness, which is obtained by adding the stripping losses at coal-dirt interfaces to the in-place covering rock thicknesses and subtracting them from the "in-place" coal thicknesses. There is no addition to the covering rock thickness in respect of the floor of the lowest seam. Thus, if v = the stripping loss at one interface, and n = the number of seams, the addition to the covering rock thickness in place = (2n - 1)v, and the subtraction from the coal thickness in place = 2nv. (DIN 21 942, 1961).

organic carbon - UK standard definition: The total carbon in the coal, less that which is present as mineral carbonate. (BS 3323 : 1978).

organic sulphur - UK standard definition: Sulphur combined in the coal substance. Note: It is the difference between the total sulphur and the sum of the pyritic sulphur and sulphate sulphur. (BS 3323 : 1978).

original reserve - US official definition: "The amount of the reserve in the ground prior to production". (USGS 1979 Proposed Revision of Bull. 1450-B, 1976).

original reserves - Taiwan official definition: Original reserves are reserves in the ground before the beginning of mine operations. In the calculation of original reserves no allowance is made for past production and losses in mining or for future losses. (Bull. geol. Surv. Taiwan, No. 10, 1959).

original resource - US official definition: "Coal in the ground prior to production".
- Criteria: "A tonnage estimate determined for coal in the ground prior to production by summing the remaining resource and cumulative depletion. A total original resource estimate is the sum of original resources determined for many mines, fields, regions, states, and the nation". (USGS 1979 Proposed Revision of Bull. 1450-B, 1976).

ortokoksowy wegiel - Polish term for a broad class of coking coals: See medium volatile coals - Polish standard classification.

outcrop verification line - : See reserves calculation method - Japanese standard procedure.

overburden ratio (overburden/coal ratio) - Reefton coalfield, NZ, procedure: This is given in cubic yards of overburden to tons of coal, and to obtain the ratio in these terms from a cross section the area of overburden should be multiplied by 1.1. For those areas where the available information warrants it, estimates should be made at several different overburden/coal ratios, <u>averaged over the whole area to be worked</u>. It is suggested that estimates to be included in the Summary of Estimates should be at the following overburden/coal ratios: bituminous coalfields 4/1; subbituminous and lignite coalfields 7/1. Estimated percentage increases or decreases for different ratios may be given. In particular areas, difficulty of disposal of overburden or the need to restore land may influence the overburden/coal ratio adopted. At the rear of any opencast the angle of the batter is assumed to be 60° in bituminous coalfields and 45° in subbituminous and lignite coalfields, the different being due to the difference in hardness of the rocks. If the angle of batter is known from experience, this should be used. (NZ geol. surv. Bull. No. 56, 1957).

overburden ratios (Verhältniszahlen Abraum:Kohle) - FRG standard procedure, for brown coal:
1) Thickness ratios:
a) The **geological thickness overburden ratio (D:K)** is calculated from the thicknesses at the covering rock and coal in place as revealed by boreholes or surface mine development.
b) The **operating thickness overburden ratio (A:K)** is obtained by substituting the "in-place" thicknesses by the "useful" thickness, which is obtained by adding the stripping losses at coal-dirt interfaces to the in-place covering rock thicknesses and subtracting them from the "in-place" coal thicknesses.
2) Quantity ratios:
a) The **operating overburden ratio (A"r" : K"r")** is obtained in every case from quantity calculations identified by index "r". This indicates how much overburden has to be removed taking the slope system required into consideration, in order to obtain a certain quantity of useful coal.
b) The **access overburden ratio (Zugangsverhältnis)** is also always got from quantity calculations. It indicates how much access to winnable coal in the surface mine has been achieved or is to be expected from an actual or predicted overburden stripping performance. All calculated volumes are to be accompanied by the following data:
i) Date for which the calculation is valid;
ii) The unit of measurement, namely the volume of overburden in m³ and the amount of coal in t;
iii) The factor used for conversion into tonnes;
iv) The minimum seam thickness used as a limiting value in the calculation;
v) The mining losses taken into account.
(DIN 21 942, 1961).

overburden subdivisions of assessed coals - Indian standard procedure: Reserve data shall be reported according to the amount of overburden on the coal in categories (1) to (5) respectively as follows: Overburden equal to one, two, three, four and five times the thickness of the seam or seams where two or more seams occur in close proximity. (Indian Standard Procedure for Coal Reserve Estimation, 1977).

P

parting - ISO definition: A lamina, for example of ankerite fusain, occurring on or at an angle to the bedding plane of a seam of coal; usually less than 3 mm thick. (ISO/R 1213/II - 1971, 1.22).

partings - : See dirt bands - Indian standard procedure.

peat lignites (ligniti torbose) - Italian definition: These have structures wherein the original matter is still visible, made up, above all, of leaves, canes, etc. They have a net calorific value of less than 1600 kcal/kg on a run-of-mine basis within the general classification of lignites. (ENI-ENEL-FINSIDER Classification Systems for Coals, 1978).

pendage - French term: See dip - CEC Analysis of Terms.

percentage deduction - NCB, UK, procedure: A percentage deduction must then (i.e. after volume/weight conversion and planar corrections) be applied to each tonnage thus calculated, as an allowance for faults, washouts and losses in working. This percentage deduction should, where possible, be based on experience in a similar geological environment of the mining method likely to be adopted. (NCB, UK, Procedure for the Assessment of Reserves, 1972).

perdite de coltivazione - Italian term: See mining losses - Italian classification.

perdite di trattamento - Italian term: See cleaning and preparation losses - Italian classification.

pertes á l'exploitation - French term: See working loss - CEC Analysis of Terms.

pertes au découpage - French term: See layout loss - CEC Analysis of Terms.

pertes prévisionnelles - French term: See planning loss - CEC Analysis of Terms. Terms.

pitch lignites (ligniti picee) - Italian classification: These show no trace either of the original matter or a woody structure. They have a net calorific value above 2200 kcal/kg (ENI-ENEL-FINSIDER Classification Systems for Coals, 1978).

planar corrections - CdF, France, procedure: The total surveyed (plan) area of a panel is corrected to take account of the gradient to give the real area (in the derivation of reserves). (CdF Reserves Classification, 1972 Model).

planar corrections - Indian standard procedure: Reserves in virgin seams or solid coal may be calculated in the case of flat seams on the basis of area, the thickness of coal beds and a correction applied in the case of seams with

P

inclination above 5°, by multiplying the figures with the secant of the dip angle. (Indian Standard Procedure for Coal Reserve Estimation, 1977).

planar corrections - NCB, UK, procedure: The acreage used will normally be the plan area but an adjustment should be made where the dip is 20° or more. (NCB, UK, Procedure for the Assessment of Reserves, 1972).

planar corrections - Ohai coalfield, NZ, procedure: The plan area has been used in all cases. Although the actual volume of coal in a given plan area for a given seam thickness increases with dip, this is offset by decreasing percentage of extraction as the dip increases. (NZ geol. Surv. Bull. No. 51, 1964).

planar corrections - Reefton coalfield, NZ, procedure: For seams dipping up to 40°, the plan area is used. Although the actual volume of coal in a given plan area for a given seam thickness increases with dip, this is offset by the decreasing percentage of extraction as the dip increases. For the seams dipping over 70°, the cross section area is used. (NZ geol. Surv. Bull. No. 56, 1957).

planar corrections - US official procedure: As the dip increases, the ground surface projection becomes inaccurate as to actual tonnage of coal underlying an area. Tonnages should be calculated in the plane of the coal bed when the dip exceeds 10° so that the tonnages for surface areas will be more precise. (USGS 1979 Proposed Revision of Bull. 1450-B, 1976).

planning loss (Verlustmenge Planung; pertes prévisionnelles) - CEC Analysis of Terms: Quantitative difference between technically recoverable reserves and planning reserves (see sketch under working loss). (CEC Assessment of Coal Reserves, 1980).

planning reserves - CdF, France, procedure: Reserves, the working of which can be planned, are obtained by adding together a1 + b1. However, a reducing coefficient may be applied to b1 in certain measures to allow for the degree of uncertainty regarding their assessment. See category "a" and "b" reserves and grade 1 reserves. (CdF Reserves Classification, 1972 Model).

planning reserves - RAG, FRG, guidelines:
i) Reserves laid out for working: Reserves of coal for which operations can be planned reliably as a result of exploration. In this case the reserves determined are present exclusively in the working areas thus defined.
ii) Reserves not laid out for working: All other coal reserves not included in reserves laid out for working. (Ruhrkohle A.G. Guidelines, 1970).

planning reserves (Planvorrat; réserves prévisionnelles) - CEC Analysis of Terms: That portion of the technically recoverable reserves which is or can be included in long-term planning. This may include tonnages which cannot be proved or located on a map. These would also be estimated tonnages, and any estimating procedure can be used. (CEC Assessment of Coal Reserves, 1980).

Planvorrat - German term: See planning reserves - CEC Analysis of Terms.

plastic range (In Audibert-Arnu test) - FRG standard definition: The softening and solidifying temperatures delimit the plastic range. (DIN 51 739, 1976, para. 3).

plomienny wegiel - Polish term for long-flame coal: See high volatile coals - Polish standard classification.

possible - : See **inferred (possible) coal** - Taiwan official classification.

possible (mogelijke/possibles) reserves - Belgian classification: Applicable in the leased areas of the Kempen/Campine coalfield to recoverable coal concerning which there is appreciable doubt about the possibility of panel layout and workability, i.e. those reserves from unworked seams known to be workable in neighbouring pits or from seams the structure of which is known only from over-working. They represent 50% of the corresponding coal in place. (Belg. Coal Min. Ind. Exec., 1963, and NVKS, 1978).

possible reserves (riserve possibili) - Italian classification: "These are that portion of reserves calculated by extrapolation of geological data in areas not systematically explored." (ENI-ENEL-FINSIDER Classification Systems for Coals, 1978).

posibles, reservas - Spanish term: See **inferred (deduced, or possible) reserves** - Colombian official classification.

possibles reserves - French term: See **possible reserves** - Belgian classification.

possible resources (mögliche Vorräte) - FRG standard procedure: For brown coal coal, these can generally only be reported as an estimate when there are individual openings in subareas of the whole region and geological investigations or data in areas which are already being worked indicate the form and extent of the deposit, or when statements concerning their presence may be made because of the particular geological conditions. The limits of error for this class are ±30% with a degree of assurance from 30% to 70%. (DIN 21 942, 1961).

potential resources (potentielle Vorräte) - FRG standard definition: For brown coal, these are all **resources** which do not fulfil present requirements for economic exploitation, but could be considered for some future use. (DIN 21 942, 1961).

potential resources (recursos potenciales) - Colombian official classification: This term, not very specific, includes a variety of mineral resources of which the possibility of economic exploitation cannot be established because of lack of geological, technological, economic or legal factors, or simply for lack of exploration or adequate infrastructure. Within this class are placed those resources classed in US geol. Surv. Bull. 1450-B as **undiscovered, unidentified, subeconomic, submarginal, paramarginal, hypothetical** and **speculative resources**, as is also a range of other class names applied locally to resources of dubious or unknown commercial value. (Publ. Geol. Esp. Ingeominas, No. 3, 1979).

potenciales, recursos - Spanish term: See **potential resources** - Colombian official classification.

potentielle Vorräte - German term: See **potential resources** - FRG standard definition.

pouvoir calorifique (spécifique) - French term: See **net (specific) calorific value** - CEC Analysis of Terms.

predicted coal reserves - Japanese standard classification: This is one of three categories into which coal reserves are generally divided, the others being **verified coal reserves** and **estimated coal reserves**. All these are further divided into types 1 and 2 depending on bed depth. (JIS M 1002 - 1978).

P

predicted coal reserves type 1 - Japanese standard definition: Coal seams within geological strata and geological structures from information other than outcrops, mines, drilling, and physical investigations in areas adjoining **estimated coal reserves type 1** are coal reserves in predicted areas, are within the present mining depth limit, and are of low assurance in comparison with **estimated coal reserves type 1.** (JIS M 1002 - 1978).

predicted coal reserves type 2 - Japanese standard definition: In terms of assurance these belong to **predicted coal reserves type 1** but the bed depth is beyond the present mining depth limit, being within the future mining depth limit. (JIS M 1002 - 1978).

presently mineable reserves (riserve attualmente coltivabili) - Italian classification. "These are all **proved, probable and possible reserves** having characteristics such as to comply with criteria of economic mining at the time of determination, such criteria varying from basin to basin. These are commonly called in situ **reserves.**" (ENI-ENEL-FINSIDER Classification Systems for Coals, 1978).

prime coking coal - : See **medium volatile coals.**

probable - : See **indicated (probable) coal** - Taiwan official classification.

probable - : See **resource group B** - Austrian guidelines.

probable reserves - : See **indicated (or probable) reserves** - Colombian official classification.

probable reserves (riserve probabili) - Italian classification: "These are that portion of reserves calculated on the basis of a spacing of observation points from 2 to 3 times the spacing which would be necessary to regard as **proved** those reserves in the same area." (ENI-ENEL-FINSIDER Classification Systems for Coals, 1978).

probable (waarschijnlijke/probables) reserves - Belgian classification: Applicable in the leased areas of the Kempen/Campine coalfield to recoverable coal from seams about which there is little information (e.g. from boreholes) and there is insufficient knowledge to determine their structure but which are known to be workable elsewhere. They represent 70% of the corresponding coal in place. (Belg. Coal Min. Exec., 1963, and NVKS, 1978).

probables, reservas - Spanish term: See **indicated (or probable) reserves** - Colombian official classification.

probables reserves - French term: See **probable reserves** - Belgian classification.

probable resources (wahrscheinliche Vorräte) - FRG standard procedure: For brown coal, if the nature of the deposit type, the thickness distribution within the occurrence and any irregularities in the deposit are broadly known from a low density borehole network and perhaps from neighbouring excavations to such an extent that an approximate calculation of resources is possible, the volumes thus calculated are classified as probable resources. The limits of error for this class are ± 20% with a degree of assurance from 70% to 90%. (DIN 21 942, 1961).

probadas, reservas - Spanish term: See **measured (or proved) reserves** - Colombian official classifiction.

P

profondeur - French term: See **depth** - CEC Analysis of Terms.

proved - : See **category "a" reserves** - CdF, France, classification.

proved - : See **measured (proved) coal** - Taiwan official classification.

proved - : See **resource group A** - Austrian guidelines.

proved reserves -: See **measured (or proved) reserves** - Colombian offical classification.

proved reserves - Indian standard procedure: In this case, the reserves are estimated from dimensions revealed in outcrops, trenches, mine workings and boreholes and the extension of the same for reasonable distance not exceeding 200 m (660 ft) on geological evidence. Where little or no exploratory work has been done, and where the outcrop exceeds 1 km (3300 ft) in length, another line drawn roughly 200 m (660 ft) in from outcrop will define a block of coal that may be regarded as proved on the basis of geological evidence. Proved reserves shall be further divided as follows:
(i)   In working collieries
         (a) Coal standing on pillars and in partings, roof and floor
         (b) Solid coal
(ii)   In closed mines
(iii) In areas covered by mining leases but not worked
(iv)   In other areas
(Indian Standard Procedure for Coal Reserve Estimation, 1977).

proved reserves (riserve provate) - Italian classification: These are that portion of reserves calculated on the basis of a spacing of observation points, varying from deposit to deposit, which permits an estimate of the quantity and quality within a margin of error of less than 10%. (ENI-ENEL-FINSIDER Classification Systems for Coals, 1978).

proved resources (sichere Vorräte) - FRG standard procedure: For brown coal, those resources for which there is a network of boreholes dense enough to permit sufficient precise assessment of the extent and form of the deposit for reliable resource calculation to be possible are considered to be proved resources. Because of differences in the formation of brown coal deposits, the borehole density required cannot be stipulated to account for every case. Borehole density is considered sufficient in each case when a substantial increase in the assurance of resource assessment can no longer be expected from any greater density in the drilling network. In particular the nature and order of magnitude of faults must be clearly enough known for it to be possible to evaluate the mining losses they would cause. The limits of error for this class are ±10% with a degree of assurance of 90%. (DIN 21 942, 1961).

proximate analysis - Indian standard definition: The analysis of coal or coke expressed in terms of percentages by weight of **moisture, volatile matter, fixed carbon** and **ash**. [IS: 1350 (Part I) - 1969].

proximate analysis - ISO definition: The analysis of coal expressed in terms of **moisture, volatile matter, ash** and **fixed carbon**. (ISO/R 1213/II - 1971, 3.22).

proximate analysis - UK standard definition: The analysis of coal expressed in terms of the **moisture, volatile matter, ash** and **fixed carbon** content. (BS 3323 : 1978).

P

proximate analysis - US standard definition: In the case of coal and coke, the determination, by prescribed methods, of **moisture, volatile matter, fixed carbon** (by difference), and **ash**.
Note: Unless otherwise specified, the term proximate analysis does not include determinations of sulphur, phosphorus or any determinations other than those named. (ANSI/ASTM D 121 - 76).

puissance - French term: See **thickness** - CEC Analysis of Terms.

puissance brute - French term: See **gross thickness** - CEC Analysis of Terms.

puissance géologique - French term: See **geological thickness** - CEC Analysis of Terms.

puissance moyenne - French term: See **average thickness** - CEC Analysis of Terms.

puissance nette - French term: See **net thickness** - CEC Analysis of Terms.

puissance pondérée - French term: See **weighted thickness** - CEC Analysis of Terms.

puissance totale - French term: See **worked thickness** - CEC Analysis of Terms.

pyritic sulphur - UK standard definition: The part of the sulphur that is present in coal in the form of pyrites or marcasite. (BS 3323 : 1978).

Q

quality - NSW, Australia, procedure: **Volatile matter** (per cent), **ash** (per cent), **calorific value** (MJ/kg), sulphur (per cent), coking properties (BSI), etc., should be recorded where applicable and, where necessary, reserves should be calculated within ranges to be specified by the calculator. The basis (i.e. **dry basis, dry ash-free, dry mineral-matter-free, as received**) used for the analyses must be stated. (NSW Code for Coal Reserves, 1979).

quality limiting criteria - Japanese standard procedure: This quality of coal obtained through calculation of coal reserves necessarily implies that it is of economically mineable scope. Coal resources are in principle not calculated for coal qualities belonging to lignite (F) in the table under **depth limiting criteria.** (JIS M 1002-1978).

quality limiting criteria - NCB, UK, procedure: From the assessment will be excluded areas of coal in which a seam is too poor in quality to work (ash and/or sulphur, etc in coal too high, or ratio of **coal** to **dirt** too low). (NCB, UK, Procedure for the Assessment of Reserves, 1972).

quality or grade - US official definition, applicable to both the **reserve** and **identified subeconomic resource** components of the classification system:"Refers to individual measurements such as heat value, **fixed carbon, moisture,** ash, sulphur, phosphorus, major, minor, and trace elements, coking properties, petrologic properties, and particular organic constituents. The individual quality elements may be aggregated in various ways to classify coal for such special purposes as metallurgical, gas, petrochemical, and blending usages". (U.S geol. Surv. Bull. 1450-B, 1976).

R

rank - : See **coal rank code** - NCB, UK, classification.

rank - ISO definition: The rank of coal is its position relative to other coals in the classification series from **brown coal** (low rank) to **anthracite** (high rank), indicating its maturity in terms of its general chemical and physical properties. (ISO/R 1213/II - 1971, 1.18).

rank - UK standard definition: The position of a coal relative to other coals in the **coalification** series from **brown coal** (low rank) to **anthracite** (high rank), indicating maturity in terms of general chemical and physical properties. Note: Higher rank is characterised by the increasing proportion of the carbon content and decreasing proportion of **volatile matter** on the **dry-mineral-matter-free** basis. (BS 3323 : 1978).

rank - US official definition, applicable to both the **reserve** and **identified subeconomic resource** components of the classification system:- "The classification of coals relative to other coals, according to their degree of metamorphism, or progressive alteration, in the natural series from **lignite** to **anthracite** (Classification of Coal by Rank, 1938, American Society for Testing Materials ASTM Designation D-388-38, p. 77-84)". (US geol. Surv. Bull. 1450-B, 1976).

rank - US official definition: The classification of coals relative to other coals, according to their degree of metamorphism, progressive alteration, or **coalification** in the natural series from **lignite** to **anthracite**. Classification is made on the basis of analysis of coal in accordance with Table 1 (of ASTM D 388). The rank of a coal can be used to infer the approximate heat value, moisture, fixed carbon and volatile matter in a coal, because the amounts of the constituents vary little within each coal rank.
- Criteria: The rank of a coal can be determined by using the instructions from the "standard specifications for classification of coals by rank" (ASTM Standards, 1975). The assignment of rank to a coal is a necessary part of proper coal classification; however, data for determining rank are often nearly lacking or far removed from the localities where the thickness of a coal is measured. In general, rank changes are gradual and occur laterally over many miles or hundreds to thousands of feet stratigraphically. Because of the lack of data, conclusions concerning rank assignments commonly must be derived from analytical determinations made on coal beds that lie some distance from where an assignment is desired. Any conclusion concerning rank where analytical data are nearly lacking must be viewed as tentative; however, if a geologist considers the geologic setting of the area sampled and the area where rank data are lacking, and extrapolates from the known to the unknown, his rank assignment probably will be correct, if his geological understanding is adequate.

R

This classification does not include a few coals, principally nonbanded varieties, which have unusual physical and chemical properties and which come within the limits of fixed carbon or calorific value of the high-volatile bituminous and subbituminous ranks. All of these coals contain less than 48 percent dry, mineral-matter-free fixed carbon or have more than 15.5 moist, mineral-matter-free British thermal units per pound. (USGS 1979 Proposed Revision of Bull. 1450-B, 1976).

raw coal - : See also **recoverable coal; run of mine.**

raw coal - ISO definition: Coal which has received no preparation other than possibly screening. (ISO/R 1213/II - 1971, 1.08).

raw coal - UK standard definition: Part of the **run-of-mine** coal separated by screening prior to preparation. (BS 3323 : 1978).

raw (roh) - FRG standard definition: The fuel as delivered is known as "raw". (DIN 51 700, 1967, para. 5.2).

recoverable coal (reserve level 2) - EMR, Canada, classification: This is derived from estimates of **mineable coal** after feasibility studies in terms of mining methods, extraction ratio and cost. It represents a mined tonnage equivalent to "run-of-mine" or "raw" coal which represents 40 per cent. to 85 per cent. of the mineable coal, depending on the mining method. Ideally, it should include estimates of cost (excluding infrastructure in remote locations) as well as tonnage (or thermal equivalent). (Canada: EMR Report EP 77 - 5, 1977).

recoverable reserves - NSW, Australia, definition and code: These consist of the amount of coal that can be physically mined from a reserve at an acceptable cost, i.e., the amount of run-of-mine coal. The term "recoverable" may be used alone or in conjunction with other terms, e.g. recoverable measured reserves, recoverable indicated reserves, etc.
For possible underground mining, recoverable reserves will be calculated for measured, and indicated reserves only.
Where the detail of knowledge warrants it, recoverable coal shall be calculated by taking measured and/or indicated reserves, and then:
1. Subtracting that coal for which there is prohibition in working such as
(a) barriers against roads, railways, lease boundaries, old workings etc;
(b) workings under stored water, rivers, swamps, tidal waters, etc;
(c) workings under roads, catchments, and other proclaimed reserves; and
(d) other restrictions imposed by lease conditions or other statutory measures.
(Note: When considering the categories listed under 1., attention should be given to prohibitions which may later apply because of surface developments or other reasons not previously stated. The assessor should state what allowance is made for such circumstances).
2. Applying a mining recovery factor supplied wherever possible by a mining engineer. In the absence of any other advice a factor of 60 per cent may be applied.
For possible open cut mining, recoverable reserves etc. recoverable reserves will be calculated for measured reserves only.
Recoverable coal shall be calculated by:
1. Subtracting that coal for which there is prohibition in working such as:
(a) barriers against roads, railways, lease boundaries, old workings, rivers etc;
(b) other restrictions imposed by lease conditions or other statutory means.
2. Applying a mining recovery factor supplied wherever possible by a mining engineer. In the absence of any other advice a factor of 90 per cent may be applied. To satisfy the high recoveries possible, the trend in thickness of the weathered zone, thickness of the seam, and other criteria critical to open

cut extraction should be accurately known. Formulation of a proving programme to enable reliable assessment of recoverable measured reserves will be necessary in accordance with the particular environment of the coal deposit. If variations occur over a wide range, calculation of reserves for a high recovery can rarely be satisfied if the points of observation are spaced at a distance greater than 100 m.
N.B. Overburden limits imposed by the calculator for both open cut and underground reserves should be stated. Overburden ratios for open cut mining should be expressed as volume (cubic metres) of non coal to mass (tonnes) of coal. (NSW Code for Coal Reserves, 1979).

recoverable reserves - Indian standard procedure: For coal in existing mines, closed mines and areas under mining leases, both in situ and **recoverable reserves** should be estimated in every case separately. In the case of working collieries the recoverable reserves shall be classified as follows:
(a) Recoverable without stowing.
(b) Recoverable with stowing.
This will also help in arriving at a recovery ratio factor applicable to the whole coalfield and the calculation of overall recoverable reserves. (Indian Standard Procedure for Coal Reserve Estimation, 1977).

recoverable reserves - Queensland, Australia, classification: These consist of coal in the ground at the date of appraisal that can be produced and utilised as coal. These reserves are calculated by subtracting estimated future losses in mining from reserves in the ground at the date of appraisal. Where the coal is to be produced to a specified grade, allowance is also made for losses in beneficiation. For all computations the following data used by the assessor are stated on the final record.
1. The percentage extraction by area after allowance has been made for pillars, barriers, and losses in mining due to any other cause.
2. The maximum height of coal to be extracted.
3. Beneficiation **recovery factor**.
4. Minimum specifications of marketable coal with respect to grade. Desirably for the purposes of this calculation, account should be taken of **measured coal** only, but it must be accepted, that for some time to come, **indicated coal first class** must be included in the calculations. Indicated coal second class and **inferred coal** should not be included. (Queensland Coal Reserves Classification, 1977).

recoverable reserves - Taiwan official classification: Recoverable reserves are reserves in the ground as of the date of appraisal that can actually be produced in the future. They are obtained by subtracting estimated future losses in mining from remaining reserves (referring to **run-of-mine coal** only, without counting the losses in coal preparation). (Bull. geol. Surv. Taiwan. No. 10, 1959).

recoverable reserves (riserve recuperabili) - Italian classification: These are in situ reserves which may be exploited from each deposit by mining systems considered most appropriate for each particular deposit. (ENI-ENEL-FINSIDER Classification Systems for Coals, 1978).

recoverable resource - US official definition:- "That part of a **resource** which will be extracted in the opinion of an estimator. The cumulative extractable parts of the total **identified** coal resources in the **economic**, marginally economic, and **subeconomic** categories which at the time of determination may be currently or potentially recoverable."
- Discussion : "Recoverable resources may be in accessed and virgin coal bodies and are generally assignable to three groups, as follows:

R

1) Resource bodies in specified thickness categories of coal that are extractable under most mining conditions;
2) Resources in specific thickness categories that underlie less than a specified minimum of overburden; and
3) Resources in coal beds that are normally too thin to mine but that may be recovered when thicker coal beds are extracted. Resources of coal under more than a specified minimum of overburden can be considered as if mining or mining plans extend to near that depth". (USGS 1979 Proposed Revision of Bull. 1450-B, 1976).

recoverable resources - UN CNR classification: These are designated by categories r-1, r-2, and r-3 subdivided, where warranted, into subcategories E, S, and M. They are the corresponding recoverable resources for each category and subcategory of in situ resources, categories R-1, etc, of economic interest for the next few decades. However, there can be no general definition of recoverability, or of the point in the mining and processing sequence at which it should be measured. (UN CNR International Classification of Mineral Resources, 1979).

recoverable resources (gewinnbare Vorräte) - FRG standard procedure: For brown coal, these are calculated by subtracting the mining losses from the calculated mining resource tonnage. Mining losses are dependent on the deposit conditions and the particular mining method employed. (DIN 21 942, 1961).

recovery factor - : See percentage deduction - NCB, UK, procedure.

recovery factor - : See recoverable reserves - Indian standard procedure.

recovery factor - Belgian procedure: In the Kempen/Campine coalfield of Belgium, recoveries are assessed on the basis of current market conditions and existing technology. Estimated recoveries are 90% of the coal in place in the case of certain reserves, 70% for probable reserves, and 50% for possible reserves. However, certain reserves are assessed only after allowances for faults, pillars and areas left between panels, whereas the lower coefficients for probable and possible reserves are applied to tonnages where fewer allowances have been made so that in many cases the recoverable tonnage is not greatly changed by an improvement in the classification of a deposit. In the case of technically workable reserves a recovery factor of 65% (based on experience in actual working in the neighbouring collieries) is applied to the calculated gross tonnage of coal in situ. (Belg. Coal Min. Ind. Exec., 1963, and NVKS, 1978).

recovery factor - CdF, France, procedure: In Charbonnage de France practice, recovery is derived as a product of a number of factors or coefficients:
i) a coefficient representing disposable output per m^3 of solid coal in place, excluding dirt, in accordance with past experience for the same seam in adjacent workings;
ii) a reducing coefficient eliminating areas which will not be worked because of geological disturbance, derived from experience of similar neighbouring measures already worked;
iii) an area reduction coefficient for planning (e.g. for an irregular area it is possible to retain only the area as an inscribed rectangle permitting efficient mechanised working); and
iv) a (dirt-free) seam thickness reducing coefficient depending on the working method (e.g. abandoning top and bottom coal, abandoning some coal as a result of working methods in steep mines, caving in, etc). (CdF Reserves Classification, 1972 Model).

recovery factor - US official definition: The percentage of total tons of coal estimated to be recoverable from a given area in relation to the total tonnage estimated to be in the **reserve base** in the ground.
- Criteria: On a national basis the estimated recovery factor for the total reserve base is 50 percent. More precise recovery factors can be computed by determining the total coal in place and the total coal recoverable in any specific locale. (US geol. Surv. Bull. 1450-B, 1976).

recovery factor (Ausbeutefaktor; factor de rendement) - CEC Analysis of Terms: Dimensionless factor applied to reserves in order to indicate the proportion of the reserves recovered (or **yield**). This factor is always $0 < f < 1$. (CEC Assessment of Coal Reserves, 1980).

recovery factor (extraction rate) - Ohai coalfield, NZ, procedure:
Opencast areas:
As in entry for Reefton coalfield, except that the loss for each stone band is increased by 1.5 ft.
Underground mining areas:
As in entry for Reefton coalfield except that it now appears unlikely that stowage, or some other method of increasing the extraction rate, will be used for many years to come, so that estimation as recoverable of more than a small proportion of the coal in thick seams is unwarranted. (NZ geol. Surv. Bull. No. 51, 1963).

recovery factor (extraction rate) - Reefton coalfield, NZ, procedure:
Opencast areas:
Some coal is lost in stripping, the amount depending on the number of partings and dip of the seam. For a seam without partings a loss of 1.5 ft. is assumed. Each stone band doubles the loss. For seams dipping over 15° an extra loss of 0.5 ft. is assumed.
Underground mining areas.
The proportion of coal recovered in underground mining varies greatly with the different seam thicknesses, being generally small where seams are over 20 ft. thick. No such thick seams have been mined in New Zealand in any manner that has recovered a large proportion of the coal, alstowing may soon be tried tried in some areas. In though view of New Zealand's dwindling reserves of better quality coals, conservation should be weighed against increased costs, and as a high a proportion won as possible. It is probable, however, that the cost factor will predominate for some years to come, and to estimate as recoverable any more than a small proportion of the coal in thick seams would present an over optimistic picture of the reserves. For a seam 5 ft thick the proportion of recoverable coal approaches 100 per cent, but the proportion recovered diminishes with increasing thickness of the seam. The maximum convenient thickness for mining is 9 ft, and at this thickness about 75 per cent is likely to be recovered. The recovery from thick seams in West Port Coalfield has been considerably less than 50 per cent, 30 per cent being a general average for seams about 35 ft thick, although a higher recovery would have been possible but for a bad mudstone roof over much of the seam. Extensive mine fires have reduced the amount recovered to less than 20 per cent in some places. The following estimated extraction rates apply to seams with good roof and floor, and should be reduced by 5 per cent where these are bad. Where a seam contains partings, only clean coal over 4 ft thick is used for the estimate.

R

Seam Thickness Ft.	Estimated Extraction per cent.	Seam Thickness Ft.	Estimated Extraction per cent.
5	95	12	60
6	90	15	50
7	85	20	40
8	80	30	30
9	75	40	25
10	70	50	20

Where it is known that some stowing method will be employed a different extraction rate should be adopted. Until such a rate is known from experience 70 per cent should be assumed. These suggested extraction rates do not preclude the use of more accurate ones where these are known. (NZ geol. Surv. Bull. No. 56, 1957).

recurso mineral - Spanish term: See **resource** - Colombian official classification.

recursos identificados - Spanish term: See **identified resources** - Colombian official classification.

recursos potenciales - Spanish term: See **potential resources** - Colombian official classification.

reduction - : See **sample reduction** - US standard definition.

remaining reserves - Taiwan official classification: Remaining reserves are reserves in the ground as of the date of appraisal and may be obtained by subtracting past production and losses from original reserves. (Bull. geol. Surv. Taiwan, No. 10, 1959).

remaining resource - US official classification: "The resource in the ground in a mine, area, field, basin, region, province, county, state, or nation after prior mining; does not include the coal lost in mining. May be divided into categories such as remaining economic, marginally economic, **subeconomic**, **measured**, **indicated**, **inferred**, **identified**, and **undiscovered resources** or other type of resources. The total remaining resource is the sum of identified and undiscovered resources as of the date of the estimate". (USGS 1979 Proposed Revision of Bull. 1450-B, 1976).

remaining resources - US official classification: Includes the sum of the **identified** and **undiscovered resources** as of the date of the estimate. (US geol. Surv. Bull. 1450-B, 1976).

rendement - French term: See **yield** - CEC Analysis of Terms.

reserva - Spanish term: See **reserve** - Colombian official classification.

reservas deducidas - Spanish term: See **inferred (deduced, or possible) reserves** -

reservas demonstradas - Spanish term: See **demonstrated reserves** - Colombian official classification.

reservas indicadas - Spanish term: See **indicated (or probalbe) reserves** - Colombian official classification.

reservas inferidas - Spanish term: See **inferred (deduced, or possible) reserves** - Colombian official classification.

reservas medidas - Spanish term: See **measured (or proved) reserves** - Colombian official classification.

reservas probables - Spanish term: See **indicated (or probable) reserves** - Colombian classification.

reservas probadas - Spanish term: See **measured (or proved) reserves** - Colombian official classification.

reservas possibles - Spanish term: See **inferred (deduced, or possible) reserves** - Colombian official classification.

reserve (reserva) - Colombian official classification: This is the portion of an identified resource of which the utilisable mineral can be legally and economically extracted at the time of evaluation of the corresponding deposit. (Publ. Geol. Esp. Ingeominas, No. 3, 1979).

reserve - US official definition: That portion of the **identified coal resource** that can be economically mined at the time of determination. The reserve is derived by applying a **recovery factor** to that component of the identified coal resource designated as the **reserve base**.
- Criteria: Reserve "includes that portion of the reserve base that can be mined at the time of classification (see recovery factor)". (US geol. Surv. Bull. 1450-B, 1976).

reserves - EMR, Canada, definition:- Refer to that portion of coal resources that has been reasonably well delineated and can be produced with current technology and delivered at competitive market prices. However, there are various ways of expressing this portion and unfortunately the term "reserves" is applied rather loosely to the various results. Ideally, reserves should be calculated from data on geology and coal quality, and reported along with production costs. Two basic criteria require to be met:
1) That infrastructure (transportation facilities, electric power, townsite, etc.) either exists or can be amortized from coal sales; and
2) That mining is permitted in the areas by government policy. If either criterion is not fulfilled, there are no reserves (e.g. there are no coal reserves in National Parks.) Reserves are further classified by **reserve levels 1, 2, 3 and 4** . (Canada: EMR Report EP 77 - 5, 1977).

reserves - Rep. South Africa, DM, definition: economically exploitable deposits of which the magnitudes are known within definite limits, and which are economically exploitable under present economic and technological conditions. For definition of 'known' see **degree of certainty**. For definition of 'limits and conditions' see **etractable coal and mineable coal in situ**. (Dept. Mines, Rept. Coal Res. SA 1975).

reserve base - US official definition:- "That portion of the **identified coal**

R

resource from which reserves are calculated".
- Criteria: Reserve base "includes beds of **bituminous** coal and **anthracite** 28 in (70 cm) or more thick and beds of **subbituminous coal** 60 in (150 cm) or more thick that occur at depths to 1000 ft (300 m). Includes also thinner and/or deeper beds that presently are being mined or for which there is evidence that they could be mined commercially at this time. Includes beds of **lignite** 60 in (150 cm) or more thick which can be surface mined - generally those that occur at depths no greater than 120 ft (40 m)". (US geol. Surv. Bull. 1450-B, 1976).

reserve base - US official definition:- "That portion of the **identified resource** delineated by physical and chemical criteria meeting minimum requirements with respect to quality, depth, thickness, rank, reliability data, and other base is the in-place portion of the **demonstrated** (measured and indicated) coal resource from which a reserve is estimated. The reserve base encompasses that part of a **resource** that has the probability, within a given planning period, of economic availability and is comprised of the resource categories economic (**reserves**), marginally economic (**marginal reserves**), and in part subeconomic (**subeconomic resources**). The term "geologic coal reserve" has been applied by others to this category but sometimes has included the inferred reserve base category".
- Criteria: "A tonnage estimate for this category of coal consists of the sum of the estimates for measured and indicated reserves, marginal reserves, and a part of the measured and indicated subeconomic resources (the coal that will be lost in mining). The reserve base is the same as the demonstrated reserve base and is preferred".
- Discussion: "Individual reserve bases, where desirable and appropriate are to be discriminated by reliability, thickness, depth, rank, overburden, chemical constituents, ash, and heat value contents, and potential usage categories. Finally, individual reserve base estimates are to be summed into totals for each coal field, basin, region, county, state, and the nation. Assignment of coal to a reserve base is controlled by physical and chemical criteria, such as, reliability of data, thickness of coal, depth of overburden, rank of coal, and knowledge of depositional patterns of coal beds and associated structural features. Changing economic, technologic, and environmental considerations do not control assignment of coal to a reserve base". (USGS 1979 Proposed Revision of Bull. 1450-B, 1976).

reserve levels - EMR, Canada, definition: These are **reserves** in levels, as variously defined below, derived from **measured** and **indicated resources**. Apart from reserve level 1, they are expressed in tons (or thermal equivalent) and, ideally, costs (excluding infrastructure costs at remote locations):
  reserve level 1   : See **mineable coal**
  reserve level 2   : See **recoverable coal**
  reserve level 3   : See **clean coal**
  reserve level 4   : See **saleable coal**. (Canada: EMR Report EP 77 - 5, 1977).

reserve, specific types - US official definition: "Reserves commonly are divided into subcategories other than those defined here-to-fore. These subcategories may be differentiated, for example: by ash, sulphur, and heat value contents; rank of coal, overburden or depth of burial; thickness, usage (metallurgical; petrochemical, and synthetic fuel types); discounted cash flow; land classification, such as State, Federal, Indian, or private ownership; reserves underlying State or National parks, monuments, forests, grasslands, military and naval reservations, alluvial valley floors, steep slopes, lakes and large rivers, and environmentally protected areas; and reserves underlying specific coalfields, districts, basins, regions, provinces, townships, quadrangles, counties, and states. The separation of coal reserves into the many different

subcategories described above and other subcategories not differentiated in the text is desirable and encouraged, but all categories should be defined so that other resource specialists and the public will not be confused".
- Criteria: "Reserve estimates are to be computed for the measured and indicated reliability categories for each rank of coal, the 0-300 ft (1-100 m) and 0-1000 ft (0-300 m) overburden categories, 28-42 in (70-105 cm) and 42 in (105 cm) thickness categories for anthracite and bituminous coal and 60-120 in (150-300 cm) and 120 in (300 cm) in thickness categories for subbituminous and lignite coal, potential mining methods, chemical constituents, such as sulfur or phosphorus content, ash content, heat value content, and usage such as metallurgical, steam, petrochemical, gasification, and liquefaction. These individual reserves estimates are calculated from the reserve base by applying an appropriate recovery factor based on economic extraction or by subtracting the coal that is estimated will be lost in mining. They are to be totalled into field, basin, region, and state estimates and a national total. Reserves include beds of bituminous coal and anthracite 28 in (70 cm) or more thick, and beds of subbituminous coal 60 in (150 cm) or more thick that occur at depths of 1000 ft (300 m); include also thinner and/or deeper beds of these coals that that are known to be economically mineable because they are currently accessed, and include beds of lignite 60 in (150 cm) or more thick which can be surface mined-generally those that occur at depths no greater than 300 ft (100 m)". (USGS 1979 Proposed Revision of Bull. 1450-B, 1976).

réserves au découpage - French term: See layout reserves - CEC Analysis of Terms.

reserves calculation method - Japanese standard procedure: Calculation of coal reserves depends on the circumstances of coal seam verification and is regulated by dividing the verification process into the following 3 cases:

(1) case where the coal seam is line-verified in mine or at outcrop (this line is known as the verification line);

(2) case where the coal seam information depends on drilling and is point-verified (this point is known as the verification point);

(3) case where respective calculated areas of coal reserves controlled by more than 2 pieces of information overlap in each of the preceding cases.

A complex set of rules governs the application of these three cases to a range

of circumstances involving variations in seam dip and in the distance between verification points and lines. The original document, which is available in English translation, explains a number of such examples, with the aid of diagrams. (JIS M 1002 - 1978).

réserves (charbon) - French term: See reserves (coal) - CEC Analysis of Terms.

reserves (coal) [Vorräte (Kohle); réserves (charbon)] - CEC Analysis of Terms: These are that part of total amount of coal in place within a designatable place which is both known to be present in the ground and whose geological and technical properties and national economic setting are sufficiently known to make it most likely to be recovered. (CEC Assessment of Coal Reserves, 1980).

réserves d'exploitation - French term: See working reserves - CEC Analysis of Terms.

réserves économiquement exploitables - French term: See economically re-

R

coverable reserves - CEC Analysis of Terms.

réserves prévisionnelles - French term: See planning reserves - CEC Analysis of Terms.

reserves subclassification - Indian standard procedure: The figures for reserves shall be classified as follows according to the rank of the coal:-

I. Anthracite

II. Bituminous
   (A) Low to medium volatile coals or coking coals: Air dried moisture up to 2% and volatile matter usually not more than 35% on unit coal basis. Sulphur less than 1%.

     By quality on the basis of the analysis of seam samples:
     Class I   - Ash not exceeding 17%
     Class II   - Ash exceeding 17% but not exceeding 24%.
     Class III - Ash exceeding 24% but not exceeding 35%.
     Class IV   - Ash exceeding 35% but not exceeding 50%.

   (B) High volatile or high moisture coals: Air dried moisture more than 2% or volatile matter usually more than 35% on unit coal basis. Sulphur less than 1%.

     (a) Semi-coking coals.
     (b) Weakly to non-coking coals.

     By quality on the basis of the analysis of seam samples:
     Class I   Ash - moisture not exceeding 19%.
     Class II   Ash - moisture exceeding 19% but not exceeding 28%.
     Class III   Ash - moisture exceeding 28% but not exceeding 40%.
     Class IV   Ash - moisture exceeding 40% but not exceeding 55%.

   (C) High Sulphur coals: Sulphur more than 1%. These occur in Assam and although some of them are semi-caking they are not suitable for metallurgical purposes.

III. Lignite
(Indian Standard Procedure for Coal Reserve Estimation, 1977).

reserves subclassification by coal class - Japanese standard classification: In terms of coal quality, coal reserves are calculated according to the classification shown in the following table:

Successive decrease in calculated distance controlled by coal class and thickness.

Coal quality	Coal thickness Class	Verified distance (1000m in case)				Distance from drilling point (dip under 30°).	
		Oblique distance (dip under 30°)		Depth (dip over 30°)			
		verified	estimated	verified	estimated	verified	estimated
Anthracite (A)	Class 1	0-500	501-1000	0-250	250-500	0-250	251-500
Bituminous coal (B, C)	Class 2	0-380	381-750	0-190	191-380	0-190	191-280
Subbituminous coal (D, E)	Class 3	0-250	251-500	0-130	131-250	0-130	131-250
Lignite (F)	Class 1	0-250	251-500	0-130	131-250	0-131	131-250
	Class 2	0-190	191-380	0-100	101-190	0-100	101-190

Note 1 unit is based on the method of counting 0.5 as a unit and discarding 0.4 and below. (JIS M 1002 - 1978).

réserves techniquement exploitables - French term: See **technically recoverable reserves** - CEC Analysis of Terms.

residual moisture - US standard definition: That moisture remaining in the sample after determining the air-dry loss. (ASTM D 3180 - 74 and ASTM D 3302 - 74).

resinite - US standard definition: Maceral derived from the resinous secretions and exudates of plant cells, occurring as discrete homogeneous bodies or clusters, individials of which are usually round, oval, or rod-like in cross section. (ANSI/ASTM D 2796 - 77).

resource distribution by access-status - : See deposit access status - Austrian guidelines.

resource group A - Austrian guidelines: Resources with an assurance of 80% to 100% are classified in this category, previously termed "Proved". (Austrian Guidelines for Coal Deposit Assessment, 1972).

resource group B - Austrian guidelines: Resources with an assurance of 50% to 80% are classified in this category, previously termed "Probable". (Austrian Guidelines for Coal Deposit Assessment, 1972).

resource group C - Austrian guidelines: Indicated resources with an assurance of less than 50% are classified in this category. (Austrian Guidelines for Coal Deposit Assessment, 1972).

resources (Lagerstättenvorräte) - FRG standard procedure: For brown coal, for all (proved, probable and possible) resource calculations, when the nature of the deposit requires it and when sufficient data are available, the calculations for the quantity of coal are to be subdivided by kinds of coal (according to use). All calculated volumes are to be accompanied by the following

R

data:
a) Date for which the calculation is valid;
b) The units of measurement, namely the volume of overburden in m^3 the amount of coal in t;
c) The factor used for conversion into tonnes (where differing coal qualities occur the conversion factors shown by density tests for each kind of coal are to be applied;
d) The minimum seam thickness used as a limiting value in the calculation;
e) The mining losses taken into account
(DIN 21 942, 1961).

resources - Rep. South Africa, DM, definition: known occurrences of coal, where the magnitudes of the deposits and their economic exploitability are either not known or are such that exploitation is not warranted at present. See also coal in situ. (Dept. Mines, Rept. Coal Res. SA 1975).

resources of future interest - EMR, Canada, definition: These are resources in coal seams that, because of less favourable combinations of thickness, quality, depth, and location are not resources of immediate interest but may become of interest in the foreseeable future. (Canada: EMR Report EP 77 - 5, 1977).

resource (recurso mineral) - Colombian official classification: A mineral resource is a natural concentration of solid, liquid or gaseous material in or on the Earth's crust, of such characteristics that its economic exploitation is actually or potentially feasible. (Publ. Geol. Esp. Ingeominas, No. 3, 1979).

resources - EMR, Canada, definition: The term coal resources refers to concentrations of coal of certain characteristics and occurring in the ground within specified limits of seam thickness and depth from surface. According to current estimates of feasibility of exploitation they are divided into two broad groups, resources of immediate interest and resources of future interest, which are further subdivided by levels of assurance of existence, into measured, indicated, inferred, and speculative resources. (Canada: EMR Report EP 77 - 5, 1977).

resources - UN CNR definition: These include all of the in situ mineral resources, classified in categories R-1, R-2, and R-3, that might be of economic interest over the foreseeable period of the next few decades. These categories are differentiated according to the level of geological assurance that can be assigned to each category. Any additional material with a lower economic potential, estimates of which would fall outside the boundaries of "resources" as here defined, should be referred to as "occurrences" and should be reported separately along with some clarification as to the derivation and meaning of the estimates.
Where warranted, each of the categories can be further subdivided into subcategory E (considered exploitable under prevailing socio-economic conditions with available technology), and subcategory S (the balance of the in situ resources that may become of interest as a result of foreseeable economic or technological changes). (UN CNR International Classification of Mineral Resources, 1979).

resources - US official definition: "Concentrations of coal in such forms that economic extraction is currently or may become feasible". (US geol. Surv. Bull. 1450-B, 1976).

resource(s), coal - US official definition: "A coal resource is a virgin or accessed naturally occurring concentration or deposit of coal in the earth's crust, in such form and amount that economic extraction is currently or may become feasible". - Commonly used subdivisions: See hypothetical, identified, indicated, inferred, measured, original, recoverable, remaining, speculative, undiscovered and resources, specific types. (USGS 1979 Proposed Revision of Bull. 1450-B, 1976).

resources of immediate interest - EMR, Canada, definition: These consist of resources in coal seams that, because of favourable combinations of thickness, quality, depth, and location, are considered to be of immediate interest for exploration or exploitation activities. Excluded from the resource estimates are Newfoundland, Manitoba and all of northern Canada north of 60°N latitude. See depth and thickness criteria for parameters corresponding to quality and location. In all areas coal beds are included that are thinner or deeper than the stated limits, if currently mined. (Canada: EMR Report EP 77 - 5, 1977).

resources, specific types - US official definition: "Resources are commonly divided into subcategories other than those defined previously. These subcategories may be differentiated by ash, sulfur, and heat value contents; rank of coal; overburden or depth of burial; steep slopes; thickness; usage (metallurgical, petro-chemical and synthetic fuel types); resources underlying specified lands owned by state, federal governments, or private interests; resources underlying State or National parks, monuments, forests, grasslands, military, Indian reservations, alluvial valley bottoms, steep slopes, lakes and large rivers, and environmentally protected areas; resources that are buried at depths below 6000 ft (1800 m) and resources underlying specific coalfields, districts, basins, regions, provinces, townships, quadrangles, counties, and states. The division of coal resources into the many different categories defined here-to-fore and other categories not defined in the text is desirable and encouraged. However, the currently defined categories may not be modified as to their definitions. Many requests for information about resources are received and are unanswerable because the scope of systems of classification used in the past was too limited. Persons and institutions classifying resources are therefore encouraged to utilize initiative in defining and developing additional classes of coal categories". (USGS 1979 Proposed Revision of Bull. 1450-B, 1976).

ressource géologique - French term: See coal in place - CEC Analysis of Terms.

restricted reserve - US official definition: "A category of reserve that is restricted from extraction by Federal, State, or local laws or regulations. A restricted reserve meets all the requirements of the reserve category, except that it is restricted from extraction. Coal assigned to this category may be either temporarily or permanently removed from an extractable category, but because laws and regulations can be repealed or changed, it should be separately distinguished and recorded as a reserve. Examples of such reserves are bodies of coal underlying: 1) archaeological and historical sites; 2) areas where endangered fauna and flora reside; 3) lands being considered as possible primitive and wilderness areas; 4) Federal or State lands that are being considered for leasing; 5) lands whose successful reclamation after mining is considered as uncertain; 6) lands whose mining would result in polluting adjacent areas; 7) alluvial valleys; 8) and so forth". (USGS 1979 Proposed Revision of Bull. 1450-B, 1976).

riserve attualmente coltivabili - Italian term: See presently mineable reserves Italian classification.

riserve commerciali - Italian term: See marketable reserves - Italian classification.

R

riserve non coltivabili attualmente - Italian term: See not presently mineable reserves - Italian classification.

riserve possibili - Italian term: See possible reserves - Italian classification.

riserve probabili - Italian term: See probable reserves - Italian classification.

riserve provate - Italian term: See proved reserves - Italian classification.

riserve recuperabili - Italian term: See recoverable reserves - Italian classification.

risorse di carbon fossile - Italian term: See coal resources - Italian classification.

risorse geologiche - Italian term: See geological resources - Italian classification.

risorse ipotetiche - Italian term: See hypothetical resources - Italian classification.

Roga index - : See roga test - UN ECE procedure.

Roga test - UN ECE procedure: The principle of the Roga method is as follows: the caking power of coal is defined by determination of the mechanical strength of crucible coke obtained by mixing 1 g of coal with a determined amount of anthracite as a diluting substance. The coke button obtained is then submitted to the drum test in a strictly prescribed manner. From the results of the drum test may be calculated the caking power of the coal tested. (The International Classification of Hard Coals by Type, 1956, app. II).

roh - : See as delivered; as received - FRG standard definition.

rough sample - FRG standard definition: Rough sample, consisting of the accumulated individual samples before further preparation. See average sample for context. (DIN 51 701, 1950, para. 2).

run-of-mine coal - UK standard definition: Coal produced by mining operations before preparation. (BS 3323 : 1978).

run-of-mine resources in place - : See volume/weight conversion - FRG standard procedure.

run of mine (r.o.m.) - ISO definition: Coal produced by mining operations, before preparation. (ISO/R 1213/II - 1971, 1.07).

run of mine (r.o.m.) - : See raw coal; recoverable coal - EMR, Canada, definition.

S

SA, sa - Abbreviation for semi-anthracite: See anthracite - Indian and US standard classifications.

safe coal reserves - Japanese standard definition: These are coal reserves which become the object of mining and for which some decrease is anticipated owing to investigation accuracy and the specific geological conditions of the coal seam within the theoretical mineable buried coal reserves of verified coal reserves type 1 (A) and of verified coal reserves type 1 (B). The ratio between these reserves and theoretical mineable buried coal reserves, expressed as a percentage, is known as safety factor. (JIS M 1002 - 1978).

saleable coal - UK standard definition: The total amount of coal output after preparation of the run-of-mine coal. It equals the total run-of-mine tonnage, less any materials discarded during preparation. (BS 3323 : 1978).

saleable coal (reserve level 4) - EMR, Canada, definition: Estimates of saleable coal represent "free-on-board" tonnages of clean coal and discounts transportation and handling losses. For some thermal coals, saleable coal may be equivalent in tonnage and cost to clean coal or even to recoverable coal. Ideally it should include estimates of costs (excluding infrastructure in remote locations) as well as tonnage (or thermal equivalent). (Canada: EMR Report EP 77 - 5, 1977).

saleable resources in place - FRG standard procedure: The run-of-mine resources in place after deduction of dirt is to be described in short in the resource calculations as the saleable resource in place. It is the product of the run-of-mine resource in place and the average yield of saleable coal as calculated from the channel samples from that seam section. The same channel samples are used in this calculation as were used in calculating the average specific gravity from the seam section. The in place resource of saleable coal is the theoretical maximum of the saleable tonnage to be achieved from the seam section. (DIN 21 941, 1953).

sample - ISO definition: A part of a population collected with the object of estimating some characteristic. It is a portion extracted from a consignment batch or unit as being representative of it with regard to the characteristic to be investigated. (ISO/R 1213/II - 1971, 2.11).

sample - US standard definition: A quantity of material taken from a larger quantity for the purpose of estimating properties or composition of the larger quantity. (ANSI/ASTM D 2234 - 76, para. 4.16).

sample division - US standard definition: The process whereby a sample is reduced in weight without change in particle size. (ANSI/ASTM D 2013 - 72).

sample reduction - US standard definition: The process whereby a sample is

S

reduced in particle size by crushing or grinding without change in weight. (ANSI/ASTM D 2013 - 72).

sampling - ISO definition: The collection of a representative portion of coal for analysis and testing. (ISO/R 1213/II - 1971, 2.03).

sclerotinite - US standard definition: A maceral having reflectance between that of fusinite and associated vitrinite and occurring as round or oval cellular bodies of varying size (20 to 300 μm) or as interlaced tissues derived from fungal remains. (ANSI/ASTM D 2796 - 77).

seam impoverishment - : See **sedimentary limitations** - Austrian guidelines.

seam thickness evaluation - Ohai coalfield, NZ, procedure: Everywhere the seam thickness is obtained from isopachs where sufficient information is available judged from knowledge of the coalfield. Quantities of measured and indicated coal are based on isopachs obtained from information on thickness from mine workings, drill-holes, and outcrops, and supplemented where necessary by an assumed rate of thinning. An average thickness over the estimate area is not assumed, the quantities being calculated for the different parts of the estimate area and added together. Where isopachs cannot be drawn, a conservative average thickness is used based on such information as is available. Such coal is classed as inferred. (NZ geol. Surv. Bull. No. 51, 1964).

sedimentary limitations - Austrian guidelines: Tectonic faults, impoverishment and the like shall be taken into consideration by using an empirically determined percentage. (Austrian Guidelines for Coal Deposit Assessment, 1972).

semi-anthracite - : See **anthracite** - Indian and US standard classifications.

semifusinite - US standard definition: The maceral that is intermediate in reflectance between fusinite and associated vitrinite, that shows plant cell wall structure with cavities generally oval or elongated in cross section, but in some specimens less well defined than in fusinite, often occurring as a transitional material between vitrinite and fusinite.
(ANSI/ASTM D 2796 - 77).

semikoksowy wegiel - Polish term for a class of coking coal: See **medium volatile coal** - Polish standard classification.

shale - ISO definition: One of the impurities associated with coal seams; this term should not be used as a general term for washery rejects.
(ISO/R 1213/II - 1971, 1.20).

shale - UK standard definition: Fine grained sedimentary rock showing marked laminations and composed essentially of minerals of the clay group but possibly including materials of high silica content. (BS 3323 : 1978).

shale band - : See **dirt band** - ISO definition.

short flame coal - UK standard definition: Coal with medium volatile matter content (approximately 20% to 25% DMMF). Note: This term is not used in the United Kingdom but is in general use in some other countries.
(BS 3323 : 1978).

sichere Vorräte - German term: See **proved resources** - FRG standard definition.

size fraction - ISO definition: The part of the sample belonging to a specified size class limited by either one or two sieve sizes. (ISO/R 1213/II - 1971, 3.17).

softening temperature (in Audibert-Arnu test) - UN ECE definition: Temperature at which the piston has moved down 0.5 mm. (The International Classification of Hard Coals by Type, 1956, app. III).

softening temperature (Erweichungstemperatur) (in Audibert-Arnu test) - FRG standard definition: The temperature at which contraction begins. (DIN 51 739, 1976, para. 3).

softening temperature (temperature de debut de ramollissement/temperatuur van beginnende verwekung) (in Audibert-Arnu test) - Belgian standard definition: The temperature at which the piston has moved down one per cent of the original length of the pencil. (NBN 831-05, 1970, para. 2).

solidifying temperature (in Audibert-Arnu test) - : See temperature of maximum dilatation - FRG standard definition.

sparking fuels - US standard definition: Within the context of this standard, fuels that do not yield a coherent cake as residue in the volatile matter determination but do evolve gaseous products at a rate sufficient to mechanically carry solid particles out of the crucible when heated at the standard rate. Such coals normally include all low-rank noncaking coals and lignites but may also include those anthracites, semianthracites, bituminous, chars and cokes that lose solid particles as described above. These are defined as "sparking fuels" because particles escaping at the higher temperatures may become incandescent and spark as they are emitted. (ANSI/ASTM D 3175 - 77).

spatial distribution of assessed coals - Indian standard procedure: The reserves of each individual seam in a sector or part of the coalfield shall be given separately. Reserve data shall be reported according to depth from the surface as follows:

1.   0 m to 150 m (0 to 500 ft)
2. 150 m to 300 m (500 to 1000 ft)
3. 300 m to 600 m (1000 to 2000 ft)
4. 600 m to 900 m (2000 to 3000 ft)
5. 900 m to 1200 m (3000 to 4000 ft)

The reserves shall be reported separately for each seam in the form tables indicating thickness taken and dip, to facilitate break up of figures qualitywise, for coal in existing mines, closed mines and areas under mining leases. (Indian Standard Procedure for Coal Reserve Estimation, 1977).

spatial distribution of assessed coals - Queensland, Australia, classification: Where reserves of coal suitable for open cut mining are reported, the upper and lower limits of overburden thickness of the ratios of overburden thickness to coal thickness should be stated. Reserves of coal suitable for underground mining should be reported in any or all of the ranges listed below. The categories applicable to the estimation should be stated.

1. Less than 150 m
2. Less than 300 m
3. Less than 450 m
4. Less than 600 m
5. Greater than 600 m
(Queensland Coal Reserves Classification, 1977).

S

spatial distribution of assessed coals - RAG, FRG, Guidelines: geological reserves are subdivided into tectonic areas by seams or beds. They are further subdivided by depth as follows:
  i) Reserves in the area of the main haulage operation, i.e. capable of being freed from the main haulage level in operation.
  ii) Reserves above the area of the main haulage level in operation.
  iii) Reserves under the area of the main haulage level in operation to a depth of 1200 m below Ordnance Datum (-1200 m NN).
  iv) Reserves between 1200 m and 1500 m below Ordnance Datum.
(Ruhrkohle A.G. Guidelines, 1970).

spatial distribution of assessed coals - Reefton coalfield, NZ, procedure: Where the adopted maximum thickness of cover exceeds 1000 ft - see depth limiting criteria - separate estimates may, if required, be made for depths to 1000 ft, from 1000 ft, to 1500 ft, 1500 ft, to 2000 ft, and 2000 ft, to 2500 ft. (NZ geol. Surv. Bull. No. 56, 1957).

specific gravity - : See volume/weight conversion - Austrian guidelines.

speculative resources - EMR, Canada, definition: These denote the precision with which given quantities of resources have been determined and estimated. They are defined as resources for which quantity estimates are based on information from a few scattered occurrences. Resources of this description are mainly in frontier areas where coal mining or exploration have not taken place. (Canada: EMR Report EP 77 - 5, 1977).

speculative resources - US official definition: "Undiscovered coal in beds that may occur either in known types of deposits in a favourable geologic setting where no discoveries have been made, or in deposits that remain to be recognized. Exploration that confirms their existence and reveals quantity and quality will permit their reclassification as reserves or identified subeconomic resources.
Criteria- "Quantitative estimates are based on geologic assumptions that undiscovered coal may occur in known types of deposits or in favorable geologic settings". (US geol. Surv. Bull. 1450-B, 1976).

speculative resources - US official definition: "An undiscovered virgin coal resource that may occur either in known types of deposits in favorable geological settings where coal deposits have not been discovered or in types of deposits as yet undiscovered for their economic potential. Exploration by surface or subsurface geologic mapping, geophysical surveying, and drilling that confirms its existence and reveals sufficient understanding so as to allow more precise estimation of quantity and quality, will permit reclassification as an identified economic or subeconomic resource , or as a reserve.
- Criteria: "As of the time of writing this report, there are no speculative resources of coal estimated for the United States. However, if it becomes desirable and necessary to make such estimates, the definition of speculative resource will be followed, the estimates will be quantitative, and the geologic evidence supporting the estimate and methods of quantification will be made publicy available". (USGS 1979 Proposed Revision of Bull. 1450-B, 1976).

splint coal - UK standard definition: Hard coal that has an even fracture and little lustre. (BS 3323 : 1978).

sporinite - US standard definition: A maceral derived from the waxy coatings (exines) of spores and pollen. (ANSI/ASTM D 2796 - 77).

standard coal seam - Japanese standard classification: Class 1 coal thickness of the thickness classes and coal qualities belonging to anthracite (A), bituminous coal (B and C), and subbituminous coal (D and E) of the Table below are called standard coal seams. This Standard takes a calculation method on this basis as standard and also regulates the calculation method used for other thin seams and coal quality (F) seams of the Table.

Coal Classification system for Reserve Subclassification by Coal Class - Japanese standard classification.

Classification		Calorific value (1) (compensated dry, ash-free basis) kcal/kg (kJ/kg)	Fuel ratio	Coking property	Remarks
Coal quality	Subdivision				
Anthracite (A)	A 1		over 4.0	non-coking	Natural coke (dandered coal) produced in igneous rock processes
	A 2				
Bituminous coal (B, C)	B 1	over 8400 (over 35160)	over 1.5	strongly coking	
	B 2		under 1.5		
	C	over 8100-8400 (over 33910-35160)		coking	
Subbituminous coal (D, E)	D	over 7800-8100 (over 32650-33910)		weakly coking	
	E	over 7300-7800 (over 30560-32650)		non-coking	
Lignite (F)	F 1	over 6800-7300 (over 29470-30560)		non-coking	
	F 2	over 5800-6800 (over 24280-29470)			

Note (1) Calorific Value (compensated dry, ash-free basis) =

$$\frac{\text{Calorific Value}}{100 - \text{ash part compensated ratio} \times \text{ash part} - \text{moisture part}}$$

Ash-free compensated ration derives from the Japan Coal Distribution Corporation method. (Reference publication: 2nd set of collected technical papers produced by the Japan coal Distr. Corp. Tech. Board and Explanation of Coal Board Standard for Calculation of Coal Reserves, April 1949). (JIS M 1002 - 1978).

steam coals - : See low volatile coals.

Steinkohle - German term: See coal (hard coal) - CEC Analysis of Terms.

Steinkohlenklassen - German term: See classes of coal - CEC Analysis of Terms.

S

subbituminous - Indian standard classification:

Class and Type	Symbol	Nature	Gross CV kcal/kg	Volatile Matter Percent (dmf)	G-K Coke Type	Moisture (60% RH) Part/100 parts Unit Coal
Sub-bituminous High Volatile	SB	Non-caking	6950 to 7500	35 to 50	A	10 to 20

Note. Based on available data, broad ranges of the properties are given. (IS: 770 - 1977).

subbituminous - US standard definition:

Class	Group	Calorific Value Limits Btu per pound (Moist * Mineral-Matter-Free-Basis) Equal or Greater Than	Less Than	Agglomerating Character
III. Sub-bituminous	1. Subbituminous A coal	10500	11500	non-agglomerating
	2. Subbituminous B coal	9500	10500	
	3. Subbituminous C coal	8300	9500	

* Moist refers to coal containing its natural inherent moisture but not including visible water on the surface of the coal.

(ANSI/ASTM D 388 - 77, Table 1).

subcategory E - UN CNR definition: Where warranted, each of the categories of resources can be further subdivided into subcategory E (considered exploitable under prevailing socio-economic conditions with available technology), and subcategory S (the balance of the in situ resources that may become of interest as a result of foreseeable economic or technologic changes). The subcategories E and S are particularly useful for subdividing resource category R-1 and perhaps category R-2, but it is not expected that category R-3 will be subdivided in practice. (UN CNR International Classification of Mineral Resources, 1979).

subcategory M - UN CNR definition: This provides for a subdivision of subcategory S to provide estimates of resources that may become exploitable in the more immediate future as a result of normal or anticipated changes in economic or technical circumstances. This subcategory also applies where warranted to the corresponding recoverable resources.
One of the more common terms that, to varying degree, has been used as equating with this category is marginal. (UN CNR International Classification of Mineral Resources, 1979).

subcategory S - UN CNR definition: "The balance of the in situ resources that is not considered of current interest but may become of interest as a result of foreseeable economic or technologic changes." This subcategory also applies where warranted to the corresponding recoverable resources.
i) The subcategories E and S are particularly useful for subdiving resource category R-1 and perhaps category R-2, but it is not expected that R-3 will generally be subdivided in practice.
ii) If desired, subcategory S may be further subdivided to provide for an estimate of resources, subcategory M, that may become exploitable in the more immediate future as a result of normal or anticipated changes in economic or technical circumstances. One of the more common terms that, to varying degree have been used as equating with this category is subeconomic.
(UN CNR International Classification of Mineral Resources, 1979).

subeconomic (bedingt abbauwürdig) - Austrian guidelines: Deposit resources no classified as economic are subeconomic if a condition of a technical or economic nature can conceivably be fulfilled, e.g.:
1. smaller or greater thicknesses than classifiable as economic, for which technical facilities are possible but not available at present;
2. roadway- or surface-protection pillars with regard to stoping method (backfilling) or mining damage costs;
3. deposit portions for the development of which special expenditure is required.
(Austrian Guidelines for Coal Deposit Assessment, 1972).

subeconomic (bedingt abbauwürdig) - FRG standard definition: - in respect of seams lying outside developed areas, refers to seams which, although technically mineable, for economic reasons either cannot be worked or can only be worked at certain times, i.e. given a suitable price and wage structure. (DIN 21 941, 1953).

subeconomic resource - US official procedure: "Tonnage estimates for this class of coal are determined by summing the estimates for measured, indicated, and inferred resources that cannot be exploited economically at the time of classification but which may become economic in the future. This category includes all measured, indicated, and inferred coal that is not assigned to an economic category. Specifically included in this category are all bituminous coal and anthracite beds in the measured and indicated categories that are 14 in (35 cm) to 28 in (70 cm) thick, and all subbituminous coal and lignite beds 30 in (75 cm) to 60 in (150 cm) thick that are less than 1000 ft (300 m) below the surface unless the coal in these beds will be recovered in the process of extracting coal from thicker beds. Also included are all beds of bituminous coal and anthracite 14 in (35 cm) or more thick and beds of subbituminous coal and lignite 30 in (75 cm) or more thick that occur at depths between 1000 ft (300 m) and 6000 ft (1800 m). Additionally included are lignite beds 30 in (75 cm) or more thick covered by 0 to 300 ft (0-100 m) of overburden, that cannot be surface mined by themselves". (USGS 1979 Proposed Revision of Bull. 1450-B, 1976).

subeconomic resources - See criteria under identified subeconomic resources - US official procedure.

subsample - US standard definition: A sample taken from another sample. (ANSI/ASTM D 2234 - 76).

sulphate sulphur - UK standard definition: The part of the total sulphur that is present in coal as sulphate. (BS 3323 : 1978).

surface mineable resources (Tagebauvorräte) - FRG standard procedure: For brown

S

coal, surface mining is extremely dependent on the deposit conditions. If it is not possible to decide whether an occurrence can be worked by surface mining techniques by comparison with workings in a similar deposit, then the occurrence must be investigated. (DIN 21 942, 1961).

surface moisture - : See free moisture (in coal) - US standard definition.

surface moisture - UK standard definition: The difference in moisture level between the moisture holding capacity and the total moisture. (BS 3323 : 1978).

swelling index - : See crucible swelling number.

T

Tagebauvorräte - German term: See **surface mineable resources - FRG standard procedure**.

technically recoverable reserves (technisch gewinnbare Vorräte; réserves techniquement exploitables) - CEC Analysis of Terms: Reserves in seams with a minimum thickness of 0.60 m of pure coal (net thickness) no deeper than 1500 m (in low-lying areas such as the Ruhrgebiet, German survey datum - 1500 m is set) and only in areas which allow the laying-out of workings are included in the calculation of the technically recoverable reserves. The minimum size of such workings depends on the local conditions. Detailed reasons should be given if other selection criteria have to be used. (CEC Assessment of Coal Reserves, 1980).

technically workable reserves - Belgian definition: Based on experience in neighbouring collieries, technically workable reserves represent 65% of the coal in place that is more than 80 cm thick in seams to a depth of 1500 m and not classified as certain, probable or possible in the area of the State Mining Concession in the Kempen/Campine coalfield of Belgium. (Belg. Coal Min. Ind. Exec., 1963, and NVKS, 1978).

technically workable resources - Cdf, France, classification: These are obtained by adding together the tonnages for categories a1 + a2 + b1 + b2 + c1 + c2. (CdF Reserves Classification, 1972 Model).

technisch gewinnbare Vorräte - German term: See **technically recoverable reserves - CEC Analysis of Terms**.

tectonic area - : See **area measurements - FRG standard procedure**.

tectonic areas - RAG, FRG, guidelines: These are outlined by:
1. Fold axes between dip variations exceeding 20 gon (18°) over a small area.
2. Faults exceeding 400 m in strike length,
    i) with throws of more than 10 m in the case of normal faults or
    ii) with displacements more than 10 m at right angles to the strata in the case of reverse faults.
3. Faults more than 50 m wide, having several parallel shear planes.
4. Zones 400 m wide bordering fault planes,
    i) in the case of 2. ii) above, unless the dip is less than 25 gon ±5 gon (22 1/2° ± 4 1/2°), and
    ii) on the north side of faults with an E - W strike where the dip is more than 25 gon (22 1/2°).
5. Trapezoidal and triangular or wedge-shaped block areas.
(Ruhrkohle A.G. Guidelines, 1970).

tectonic disturbance - Austrian guidelines: In coal resource assessment, tectonic faults, impoverishment and the like shall be taken into consideration by using an empirically determined percentage. (Austrian Guidelines for Coal Deposit Assessment, 1972).

tectonic disturbance - Indian standard procedure: In arriving at figures of reserves the dislocation caused by faults should be taken into account and their effect eliminated while demarcating the sectors under different categories for the purpose of calculation. (Indian Standard Procedure for Coal Reserve Estimation, 1977).

tectonic disturbance - NCB, UK, procedure: From the assessment will be excluded areas of coal in which a seam is too tectonically disturbed to work. (NCB, UK, Procedure for the Assessment of Reserves, 1972).

temperature de contraction maximale - French term for maximum contraction temperature : See inflation temperature - Belgian standard definition.

temperature de debut de ramollissement - French term: See softening temperature - Belgian standard definition.

temperature de dilatation maximale - French term: See temperature of maximum dilatation - Belgian standard and UN ECE definitions.

temperature of maximum dilatation (temperature de dilatation maximale/temperatuur van maximale dilatatie) (in Audibert-Arnu test) - Belgian standard definition: Temperature at which the piston reaches its highest position. (NBN 831 - 05, 1970, para. 2).

temperature of maximum dilatation (in Audibert-Arnu test)- UN ECE definition: Temperature at which the piston reaches its highest point. (The International Classification of Hard Coals by Type, 1956, app. III).

temperature of maximum dilatation (Verfestigungstemperatur) (in Audibert-Arnu test) - FRG standard definition: The temperature at which the process of dilatation is completed is (conventionally known as) the solidifying temperature. If no dilatation takes place, then the temperature at completion of the process of contraction is considered to be the solidifying temperature. (DIN 51 739, 1976, para. 3).

temperatuur van beginnende verweking - Flemish term: See softening temperature - Belgian standard definition.

temperatuur van maximale dilatatie - Flemish term: See temperature of maximum dilatation - Belgian standard definition.

Teufe - German term: See depth - CEC Analayis of Terms.

theoretical buried coal reserves - Japanese standard classification: Theoretically calculated buried coal reserves (hereinafter called theoretical buried coal reserves) are divided into theoretical mineable buried coal reserves and theoretical unmineable buried coal reserves according to mineability. (JIS M 1002 - 1978).

theoretical mineable buried coal reserves - Japanese standard definition: These are the buried coal reserves of a mineable area. (JIS M 1002 - 1978).

theoretical unmineable buried coal reserves - Japanese standard definition: These are the buried coal reserves of an unmineable area on sterilised land and in other remote localities, etc. between such areas and mining areas.

T

(JIS M 1002 - 1978).

thickness (Mächtigkeit; puissance) - CEC Analyis of Terms: Shortest distance (normal to the stratification) between two limiting planes of a body of rock at a particular point (true thickness). In coal mining: The distance, normal to the stratification, between the limiting planes of a stratum. (CEC Assessment of Coal Reserves, 1980).

thickness classes - Japanese standard classification: Coal seams of over 30 cm coal thickness are classified into 3 classes as follows:

Class	Coal Thickness, cm
Class 1	over 100
Class 2	over 60 but under 100
Class 3	over 30 but under 60

(JIS M 1002 - 1978)

thickness classes - Queensland, Australia, classification: Categories of recoverable reserves are also to be classified according to the thickness of coal and thickness of overburden. The ranges of coal thickness selected are
1. 1 m to 1.5 m;
2. Greater than 1.5 m.
The upper limit of thickness should be stated in respect of range (2). The sub-total of reserves in range (1) should be computed and reported separately from range (2) except for small areas where the thickness of the workable coal falls within both ranges. (Queensland Coal Reserves Classification, 1977).

thickness classes - Taiwan official procedure: The following thickness categories for reporting coal reserves are considered most desirable in Taiwan: A) 60 cm - 1 m; B) 40 cm - 60 cm; C) 25 cm - 40 cm; D) 20 cm 25 cm (optional, but not to be included in the summary totals unless specially noted). Very few coal beds in Taiwan have a thickness of more than one metre. Coal beds of more than one metre thick are included in the (A) category in estimating coal reserves. (Bull. geol. Surv. Taiwan, No. 10, 1959).

thickness evaluation - FRG standard procedure: Seam parts which are to have their coal resources calculated are divided into sections of which one can expect there to be uniform development from the probable variation as shown on the map of seam properties. For each one of these sections the seam area is taken from the tectonic map and the expected average "mined thickness" from the structural map. (DIN 21 941, 1953).

thickness evaluation - Reefton coalfield, NZ, procedure: In any area the best means of obtaining the seam thickness is by drawing isopachs, but this can be done only where sufficient information is available, this being judged from a knowledge of each particular coalfield. In general, if the information is sufficient to draw isopachs, it is also sufficient to classify the coal as measured coal or indicated coal. When isopachs are used, individual totals have to be calculated for different parts of an estimate area and added together. Where isopachs are not warranted an average thickness must be used based on such information as is available. Where indicated coal is based on an assumed rate of thinning to the workable minimum of 4 ft, the seam thickness is taken to be the thickness of coal at the limit of definite information, plus four, the total being divided by two. For inferred coal conservative figure for the average seam thickness should be adopted. (NZ geol. Surv. Bull. No. 56, 1957).

thickness evaluation - Taiwan official procedure: For coal beds known to be characteristically irregular and lenticular, a strict application of the conventional method in estimating the average coal thickness is not practical. The reserve estimate for such irregular coal beds is much more subject to arbitrary evaluation than could be justified by geologic information. The average thickness range of the coal bed may be obtained by the following theoretical formula:

$$Ts = \frac{\frac{(T_1+T_2)L_1}{2} + \frac{(T_2+T_3)L_2}{2} + \frac{(T_3+T_4)L_3}{2} + \ldots}{L_1 + L_2 + L_3 + \ldots}$$

Where:
Ts....... average thickness of the coal bed in question;
T1,T2,T3..thickness of succeeding coal outcrops along the coal bed;
L1,L2,L3..distance between two neighbouring coal outcrops along the coal bed.
(Bull. geol. Surv. Taiwan, No. 10, 1959).

thickness limiting criteria - : See thickness classes - Queensland, Australia classification.

thickness limiting criteria - Austrian guidelines:
1. Minimum seam thickness: Deposit resources up to a minimum seam thickness of 1.0 m shall be recorded; lower thicknesses shall be neglected.
2. Dirt bands: Dirt bands and partings of a thickness below 10 cm shall not be considered; known dirt bands and partings with a thickness of 10 cm and above shall be deducted from the total thickness. (Austrian Guidelines for Coal Deposit Assessment, 1972).

thickness limiting criteria - Belgian procedure: The assessment of coal in the Kempen/Campine coalfield of Belgium excludes seams with a thickness of coal less than 80 cm. (Belg. Coal Min. Ind. Exec., 1963, and NVKS, 1978).

thickness limiting criteria - CdF, France, procedure: In the Bassin du Nord and Pas-de-Calais, seams with a total thickness of coal (i.e., excluding dirt bands) of 60 cm or more are included in the calculation of reserves. The minimum seam thickness is greater in the other coalfields. (CdF Reserves Classification, 1972 Model).

thickness limiting criteria - Colombia: The thickness criteria considered economic by INGEOMINAS, by resource class and degree of metamorphism, are:

Resource Class	Degree of Metamorphism	Thickness m.
Total, including potential resources.	Anthracite and bituminous coal	0.35 or more
	Subbituminous coal and lignite	0.75 or more
Identified resources.	Anthracite and bituminous coal	0.35 or more
	Subbituminous coal	0.75 or more
Reserve base (geological or "in situ"); also recoverable (or exploitable) reserves (including the non-recoverable part of the reserve base).	Anthracite and bituminous coal	0.70 or more
	Subbituminous coal	1.50 or more
	Lignite	1.50 or more
Subeconomic (potential) resources	Anthracite	0.35 to 0.70
	Bituminous coal	0.35 or more
	Subbituminous coal	0.75 or more
	Lignite	0.75 or more

(Publ. Geol. Esp. Ingeominas, No. 3, 1979).

thickness limiting criteria - EMR, Canada, procedure: For resources of immediate interest, the minimum thickness of coal of all ranks in the Cordillera and Plains regions is 5 ft; in the Maritimes it is 3 ft for onshore coals and 5 ft for subsea coals to a distance of 5 miles offshore. For resources of future interest the minimum thickness is 5 ft in the Cordillera and for subsea coals off the Maritimes; the minimum thickness is 3 ft in the Plains and onshore Maritimes regions. (Canada: EMR Report EP 77 - 5, 1977).

thickness limiting criteria - Indian standard procedure: The average thickness of coal seams shall be calculated and stated in centimetres (and feet). Partings greater than 5 cm (2 in) in thickness, shall be excluded in calculating reserves. The burnt out portions of coal and jhama shall be excluded while taking thickness of the seam for the purpose of calculation. The thickness range for individual seams for the calculation of reserves shall be as follows:-
(a) 0.5 to 1.5 m (1.5 to 4.5 ft)
(b) 1.5 to 3.5 m (4.5 to 10.5 ft)
(c) 3.5 to 5 m (10.5 to 15 ft)
(d) 5 to 10 m (15 to 30 ft)
(e) above 10 m (above 30 ft)
(Indian Standard Procedure for Coal Reserve Estimation, 1977).

thickness limiting criteria - Japanese standard procedure: Coal reserves with a coal part thickness in the coal seam (hereinafter called coal thickness) of under 30 cm are not calculated. This restriction does not, however, extend to what is economically mineable. Coal reserves with a coal thickness percentage of under 50% within the coal seam mining part thickness (hereinafter called mining thickness) are in principle not calculated. This restriction does not, however, extend to what is economically mineable. Coal reserves are in principle not calculated for class 3 coal thickness of the thickness classes. (JIS M 1002 - 1978).

thickness limiting criteria - NCB, UK, procedure: From the assessment will be excluded areas of coal in which a seam is too thin to work (seams under 60 cm thick to be included only when of special quality or customarily worked). (NCB, UK, Procedure for the Assessment of Reserves, 1972).

thickness limiting criteria - NSW, Australia, procedure: In all cases the seam thicknesses or seam thickness ranges used in the reserve calculations should be stated. (NSW Code for Coal Reserves, 1979).

thickness limiting criteria - Ohai coalfield, NZ, procedure: In underground minning estimate areas no seam or area of a seam less than 5 ft thick is included, and no thicker seam with stone bands, or area of such a seam, is included unless it contains a minimum of 5 ft of coal without any stone. For opencast estimate areas the corresponding figure is 4 ft. Workings are unlikely to extend into coal less than 5 ft thick unless thicker coal is known beyond, nor are they likely to be started in coal less than 5 ft thick unless there is promise of the coal thickening. Small mining parties may drive in coal thinner than 5 ft in the hope that the coal will thicken inwards, but such areas are insufficiently well known for the recoverable coal to be estimated. (NZ geol. Surv. Bull. No. 51, 1964).

thickness limiting criteria - RAG, FRG, guidelines: All reserves in seams or beds with a thickness of coal alone of 60 cm or more are to be included in the calculation of geological reserves. Coal seams or beds with a thickness of coal alone of less than 60 cm are also to be included if currently worked. (Ruhrkohle A.G. Guidelines, 1970).

thickness limiting criteria - US official procedure: Applicable only to those coal bodies that are or will be economically extractable by underground, surface, and/or in situ methods, coal thinner than 14 in (35 cm) (anthracite and bituminous) and 30 in (75 cm) (subbituminous and lignite) is excluded from all Department of the Interior resource and reserve estimates from January 1, 1975. These thinner coals will be considered at a later date. To qualify for classification within the reserve base coals have to be at least twice the thicknesses quoted above against ranks. However, identified resources and the reserve base include some beds that are thinner than the general criteria permit, but that are being mined or are judged to be mineable commercially at this time. (US geol. Surv. Bull. 1450-B, 1976).

thickness limiting criteria - Reefton coalfield, NZ, procedure: (1) No seam or area of a seam less than 4 ft thick is included, and no thicker seam or area of a seam with stone bands is included unless it contains a minimum of 4 ft of coal without any stone. (2) No unworked seam is included unless the thickness is 5ft or more at at least two places not less than 4 chains apart. Few workings are likely to extend into coal less than 4 ft thick unless thicker coal is known beyond. Five feet is adopted as the minimum outcrop or drillhole thickness in which mining will be started, although where the structure is good a seam 4 ft. thick might attract a co-operative mining party. Seams over 4 ft thick containing not more than one stone band less than 3 in thick may be included provided that a note is made against the estimate. (NZ geol. Surv. Bull. No. 56, 1957).

thickness limiting criteria - Taiwan official procedure: Coal estimates are made on the basis of a 25 cm cut-off as the minimum thickness of all the coals in Taiwan. Reserves in thinner coal beds (in the range of 20 to 25 cm) may be reported for any area where the mining of such beds is economically feasible or the coal is of excellent quality such as high-grade metallurgical coking coal. However, the reserves of these thinner beds shall not be included in the summary total of the coal reserves in Taiwan, but they might be indicated separately with specific reference to their potential value. (Bull. geol. Surv. Taiwan, No. 10, 1959).

Tiefbauvorräte - : See deep mineable resources - FRG standard definition.

tonnage (Menge) - CEC Analysis of Terms: An indication of the weight of a substance. The pond megapond is the physical unit of measurement. Internationally the metric ton (t) is still used to describe coal deposits. The following conversion rates apply:
$$1 t = 1.1023 \text{ short tons} = 0.9482 \text{ long tons}$$
(CEC Assessment of Coal Reserves, 1980).

total ash - ISO definition: Residue of the mineral matter obtained by incinerating coal under defined conditions. (ISO/R 1213/II - 1971, 3.02).

total ash - UK standard definition: The inorganic residue obtained by incinerating coal under defined conditions. Note: Total ash is the sum of the extraneous ash and the inherent ash. (BS 3323 : 1978).

total moisture - ISO definition: The moisture in the coal as sampled, and removable under standardized conditions which are defined in ISO Recommendation R 589, Determination of total moisture. (ISO/R 1213/II - 1971, 3.06).

total moisture - UK standard definition: The moisture associated with the coal

as sampled. (BS 3323 : 1978).

total moisture - US standard definition: Total moisture in coal is that moisture determined as the loss in weight in an air atmosphere under rigidly controlled conditions of temperature, time and air flow as established in Method D 3302. (ANSI/ASTM D 121 - 76).

total original resources - US official definition: - includes the sum of the remaining resources and cumulative production as of the date of the estimate. (US geol. Surv. Bull. 1450-B, 1976).

tranche densimétrique - French term: See density fraction - CEC Analysis of Terms.

trend - ISO definition: A material is said to show trend if the average value of the characteristic varies unidirectionally over some interval of time, mass or space in the material. (ISO/R 1213/II - 1971, 1.27).

tv - Austrian guidelines: Abbreviation for tonnes saleable coal, in which terms only the sums of resource groups A and B , as well as the total economically mineable quantity thereof shall be stated in passing data to third parties. (Austrian Guidelines for Coal Deposit Assessment, 1972).

U

ultimate analysis - Indian standard definition: For the purpose of this standard, the term 'ultimate analysis' shall mean the analysis of coal in terms of its carbon, hydrogen, nitrogen, sulphur and oxygen contents. But for the basis of ultimate analysis or any part thereof, moisture and ash content shall also have to be determined concurrently. (IS: 1350, Part IV, 1975, Sec 2).

ultimate analysis - ISO definition: The analysis of coal expressed in terms of its carbon, hydrogen, nitrogen, sulphur and oxygen content.
(ISO/R 1213/II - 1971, 3.21).

ultimate analysis - UK standard defintion: The analysis of a coal expressed in terms of the carbon, hydrogen, nitrogen, sulphur and oxygen content.
(BS 3323 : 1978).

ultimate analysis - US standard definition: In the case of coal and coke, the determination of carbon and hydrogen in the material, as found in the gaseous products of its complete combustion, the determination of sulfur, nitrogen and ash in the material as a whole, and the estimation of oxygen by difference.
Note 1 - The determination of phosphorus is not by definition a part of the ultimate analysis of coal or coke, but may be specified when desired.
Note 2 - When the analysis is made on an undried sample, part of the hydrogen and oxygen as determined is present in the free moisture accompanying the coal. Therefore, in comparing coals on the basis of their ultimate analysis, it is advisable always to state the analysis on both the "as-received" and "dry" bases.
Note 3 - In as much as some coals contain mineral carbonates, and practically all contain clay or shale containing combined water, a part of the carbon, hydrogen, and oxygen found in the products of combustion may arise from these mineral components. (ANSI/ASTM D 121 - 76).

unbauwürdig - German term: See uneconomic - FRG standard definition.

unclassified reserves - Indian standard procedure: Where no data are available or where reliable data are lacking, the reserves should be placed under a category of 'unclassified' reserves and for the calculation of the quantity a specific gravity of 1.5 may be used. These will be split up into various categories and classes as and when reliable data becomes available. (Indian Standard Procedure for Coal Reserve Estimation, 1977).

unclassified reserves - NCB, UK, definition: Unclassified reserves, are those about which there is:-
(i) insufficient geological information for them to be classified as economically workable, but which are in a geological environment from which it can be inferred that, with further proving, a proportion may achieve higher status;
(ii) sufficient geological information to indicate that they are not economic-

U

    ally workable under present conditions, but might become so with beneficial changes in mining methods and/or marketing requirements.
Geological conditions determining classification:
In the unclassified reserves the limited data available of the sedimentary environment may be insufficient to be certain that at least some of the economically accessible areas of the seam will, with further proving, become economic and therefore classifiable. (NCB, UK, Procedure for the Assessment of Reserves, 1972).

undiscovered resources - US official definition: Unspecified bodies of coal surmised to exist on the basis of broad geologic knowledge and theory.
- Criteria: Include beds of **bituminous** coal and **anthracite** 14 in (35 cm) or more thick and beds of **subbituminous** coal and lignite 30 in (75 cm) or more thick that are presumed to occur in unmapped and unexplored areas to depths of 6000 ft (1800 m). (US geol. Surv. Bull. 1450-B, 1976).

undiscovered resources - US official definition: "An unspecified virgin deposit of coal believed to exist on the basis of geologic knowledge and theory, projection of data for long distances, geologic evidence derived from mapping and geophysical surveys, or identification of similar rocks that elsewhere are generally coal-bearing".
- Criteria: "A tonnage estimate for this category of resources is estimated by summing the tonnage estimates for coal assigned to the hypothetical and speculative resource categories. Included are unknown resources of **bituminous** coal and **anthracite** in beds 14 in (35 cm) or more thick and unknown resources of **subbituminous** coal and lignite in beds 30 in (75 cm) or more thick presumed to occur in unmapped or unexplored areas to depths of 6000 ft (1800 m)". (USGS 1979 Proposed Revision of Bull. 1450-B, 1976).

uneconomic (nicht abbauwürdig) - Austrian guidelines: Portions of a deposit not classified as economic or subeconomic are classified uneconomic, e.g. if there are:
(1) smaller or greater thicknesses than included in the subeconomic class, and
(2) protective pillars of a special kind, e.g. for headwater protection areas.
(Austrian Guidelines for Coal Deposit Assessment, 1972).

uneconomic (unbauwürdig) - FRG standard definition: In respect of seams lying outside developed areas, refers to all seams, irrespective of thickness which are neither economic nor subeconomic. (DIN 21 941, 1953).

Untersuchungsgrad - German term: See degree of exploration - CEC Analysis of Terms.

V

Verfestigungstemperatur - German term for solidifying temperature: See temperature of maximum dilatation - FRG standard definition.

Verhältniszahlen Abraum: Kohle - German term: See overburden ratios - FRG standard procedure.

verification line - : See reserves calculation method - Japanese standard procedure.

verification point - : See reserves calculation method - Japanese standard procedure .

verified coal reserves - Japanese standard classification: This is one of three categories into which coal reserves are generally divided, the others being estimated coal reserves and predicted coal reserves. All these are further divided into types 1 and 2 depending on bed depth. Verified coal reserves type 1 are, however, again divided into A and B. (JIS M 1002 - 1978).

verified coal reserves type 1 (A) - Japanese standard definition: These are coal reserves of a range where over 2 coal seams have been verified in one-way worked drifts, mines etc. in a developed area as a whole. Where the length of the verified faces is over 1 km, however, the area with the part exceeding 1 km is not calculated as A. (JIS M 1002 - 1978).

verified coal reserves type 1 (B) - Japanese standard definition: Coal seams from outcrop, mine, and drilling information in areas adjoining verified coal reserves type 1 (A) in undeveloped areas or developed areas are coal reserves of areas regarded as assured, are within the present mining limit depth, and have fairly low assurance in comparison with verified coal reserves type 1 (A). (JIS M 1002 - 1978).

verified coal reserves type 2 - Japanese standard definition: In terms of assurance these belong to verified coal reserves type 1 (B) but the bed depth is beyond the present mining method depth limit, being within the future mining depth limit. (JIS M 1002 - 1978).

verified distance - : See reserves calculation method - Japanese standard procedure.

Verlustmenge Abbau - German term: See working loss - CEC Analysis of Terms.

Verlustmengenfaktor - German term: See loss factor - CEC Analysis of Terms.

Verlustmenge Planung - German term: See planning loss - CEC Analysis of Terms.

Verlustmenge Zuschnitt - German term: See layout loss - CEC Analysis of Terms.

V

vet - Flemish term meaning "fat": See **fat coal** - Belgian standard definition.

vitrain - UK standard definition: A brilliant black coal that is glassy in appearance and brittle. (BS 3323 : 1978).

vitrain - US standard definition: Shiny black bands, thicker than 0.5 mm, of subbituminous and higher rank banded coal. (ANSI/ASM D 2796 - 77).

vitrinite - UK standard definition: The group of macerals derived from the original cell tissues of the plant material. It is normally the major maceral component of coal. (BS 3323 : 1978).

vitrinite - US standard definition: The maceral and maceral group composing all or almost all of the vitrain and like material occurring in attrital coal, as the component of reflectance intermediate between those of exinite and inertinite. (ANSI/ASTM D 2796 - 77).

vluchtige bestanddelen - Flemish term : See **volatile matter** - Belgian standard definition.

v.m. - Abbreviation: See **volatile matter**.

volatile matter - Australian standard definition: Material other than moisture which is driven off when coal is heated under standard conditions. The basis used is dry, mineral-matter-free (Est) (for formula, see dry-mineral-matter-free). (AS K 184 - 1969, para. 3.5).

volatile matter - CdF, France, classification: The **volatile matter class** (in %) (AFNOR Standard), determined by the laboratory from extracted coal which has been treated in the washery, is taken for the cleanest commercial fraction, usually from coal with 8%-10% ash content (and registered with the reserves with which it is associated). (CdF Reserves Classification, 1972 Model).

volatile matter - Indian standard definition: "Total loss in weight minus the moisture when coal or coke is heated under specified conditions". (IS: 1350, Part I, 1969).

volatile matter - ISO definition: The loss in mass, corrected for moisture when coal is heated out of contact with air under standarized conditions, which are defined in ISO Recommendation R 562 - Determination of the volatile matter of hard coal and of coke. (ISO/R 1213/II - 1971, 3.11).

volatile matter - UK standard definition: The loss in mass, less that due to moisture, when coal is heated under standard conditions and out of contact with air. (BS 3323 : 1978)).

volatile matter - US standard definition: Those products, exclusive of moisture, given off by a material as gas or vapor, determined by definite prescribed methods which may vary according to the nature of the material.
Note - In the case of coal and coke, the methods employed shall be those prescribed in ASTM Method D 3175, Test for Volatile Matter in the Analysis Sample of Coal and Coke. (ANSI/ASTM D 121 - 76).

volatile matter (flüchtigen Bestandteilen) - FRG standard definition: Volatile matter consists of the products of decomposition of organic fuel matter given off when solid fuels are heated to 900°C in the absence of air under conditions established by convention. (DIN 51 720 - 1978).

volatile matter (matiéres volatiles/vluchtige bestanddelen) - Belgian standard definition: Volatile matter content is the loss of weight, expressed as a percentage, which occurs on heating solid fuel in the absence of air in specified conditions, after allowing for the loss of weight due to the evaporation of moisture contained in the sample for analysis. (NBN 831-01 - 1970).

volume/weight conversion - Austrian guidelines: The specific gravity of the coal for sale, determined as an average specific gravity of the individual grades, and the grade key shall be used for calculating the tonnage from the geometrically determined deposit volume. This takes account of the processing recovery as well. (Austrian Guidelines for Coal Deposit Assessment, 1972).

volume/weight conversion - Belgian procedure: In the assessment of coal in the Kempen/Campine coalfield of Belgium the tonnage of coal in place is assessed by calculating the workable area of a seam x coal thickness (without dirt) x specific gravity of 1.35. (Belg. Coal Min. Ind. Exec., 1963, and NVKS, 1978).

volume/weight conversion - FRG standard procedure: The average specific gravity of the seam content to be mined in a section is got by averaging the specific gravities of samples which have been taken from channels in that section or in its vicinity, which have encountered the seam in its expected average formation and which all belong to the leaves constituting the "mined section" and only these. The product of seam area (F) in square metres, seam thickness (m) in centimetres and average specific gravity (wf) in t/m^3 divided by 100 gives the resources in place of run-of-mine coal (wf) in tonnes. In more precise calculations, i.e. when calculating the output from one operating point, attention should be paid to the fact that the prepared products have varying dirt and moisture contents. If the archive documents are not sufficient for the calculation of the various resource figures, and if no other suitable values for the averages to be applied for yield and specific gravity, then the following equivalent given by Lehmann (Glückauf 77, 1941, p 213 et seq.) are to be used:
1 cubic metre of seam = 1.39 tonnes run-of-mine coal in place
                      = 1.18 tonnes saleable coal in place
                      = 1.25 tonnes run-of-mine production
                      = 1.00 tonnes saleable production.
(DIN 21 941, 1953).

volume/weight conversion - Indian standard procedure: Where reliable data are available, the following average specific gravity of each class within each category should be used:
(A) Low to medium volatile coals or coking coals
    (a) Class I   - 1.42
    (b) Class II   - 1.47
    (c) Class III   - 1.57
    (d) Class IV   - 1.70
(B) High volatile or high moisture coals
    (a) Class I   - 1.40
    (b) Class II   - 1.45
    (c) Class III   - 1.55
    (d) Class IV   - 1.70
(C) High sulphur coals
    (a) 0 to 5%   - 1.30
    (b) 5 to 10%   - 1.34
    (c) 10 to 15%   - 1.38
See reserves, subclassifications for class qualities, based on seam samples. Where no data are available or where reliable data are lacking, the reserves should be placed under a category of unclassified reserves and for the calculation of the quantity a specific gravity of 1.5 may be used. (Indian Standard

Procedure for Coal Reserve Estimation, 1977).

volume/weight conversion - NCB, UK, procedure: The following standard volume/ tonnage conversion factors, based on the specific gravity of saleable large coal of moderate ash are to be used:-
i) 13737 tonnes per metre-hectare for anthracite;
ii) 13179 tonnes per metre-hectare for all other coals.
The conversion factor should be applied to the product of the gross acreage of the area of reserves and the total thickness of coal (i.e. inclusive of inferior coal but not dirt) in the probable mined section of the seam within that area. (Coal and **inferior coal** comprise material with ash not exceeding 40.0%) (NCB, UK, Procedure for the Assessment of Reserves, 1972).

volume/weight conversion - NSW, Australia, procedure: In the absence of other precise data it may be assumed that 1 m/km2 will contain 1.4×10^6 t of coal. (NSW Code for Coal Reserves, 1979).

volume/weight conversion - Ohai coalfield, NZ, procedure: In the absence of sufficient determinations for coals of the Ohai Coalfield, the weight/volume factor of 1500 tons/foot-acre used in estimating reserves in the Greymouth (1952) and Reefton (1957) coalfields, is used. (NZ geol. Surv. Bull. No. 51, 1964).

volume/weight conversion - RAG, FRG, guidelines: The relative densities of coal and dirt are: Coal 1.45; Dirt 2.3. (Ruhrkohle A.G. Guidelines, 1970).

volume/weight conversion - Reefton coalfield, NZ, procedure: In the absence of sufficient determinations of specific gravities of coal from different districts, the weight/volume factor of 1500 tons/foot-acre, used (1952) in estimating quantities in Greymouth Coalfield, should be used. At Greymouth the weight /volume factor based on determinations of specific gravities ranges from 1421 to 1591 tons/foot-acre, no relation with ash content, type, or rank of coal being apparent. From such information as is available a figure of 1500 tons-foot acre is unlikely to introduce significant errors in estimates for all New Zealand coals, except possibly for low-rank lignites for which further specific gravity determinations are required. The overburden/coal ratio is given in cubic yards of overburden to tons of coal, and to obtain the ratio in these terms from a cross section the area of overburden should be multiplied by 1.1. (NZ geol. Surv. Bull. No. 56, 1957).

volume/weight conversion - Taiwan official procedure: Since most of the Taiwan coals are of high volatile bituminous to subbituminous rank, a value of 1.3 is used as the common specific gravity for all the coals in computing coal reserves when precise data are not available. This value conforms closely to the average of the recorded specific gravities of the coal in those ranks. (Bull. geol. Surv. Taiwan, No. 10, 1959).

Vorräte (Kohle) - German term: See **reserves (coal)** - CEC Analysis of Terms.

W

waarschijnlijke - Flemish term : See probable reserves Belgian definition.

waf - Abbreviation for wasser- und aschefrei : See dry, ash-free... - FRG standard definition.

wahrscheinliche Vorräte : See probable resources, - FRG standard definition.

washed coal - UK standard definition: Coal from which impurities have been removed by treatment in liquid media. (BS 3323 : 1978).

wasserfrei - German term for moisture-free : See dry - FRG standard definition.

wasser und aschefrei (waf) - German term for moisture and ash-free : See dry, ash-free - FRG standard definition.

wasser und mineralstofffrei - German term for moisture and mineral-matter-free : See dry, mineral-matter-free - FRG standard definition.

water of constitution - UK standard definition: Water, chemically bound to the mineral matter that remains after the determination of total moisture. (BS: 3323 : 1978).

wegiel antracytowy - Polish term for anthracitic coal : - See low volatile coals - Polish standard classification.

wegiel chudy - Polish term for lean coal : - See medium volatile coals - Polish standard classification.

wegiel gazowo-koksowy - Polish term for a class of coals embracing gas coal and coking coal : - See high volatile coals - Polish standard classification.

wegiel gazowo-plomienny - Polish term for long-flame gas coal : See high volatile coals - Polish standard classification.

wegiel gazowy - Polish term for gas coal : See high volatile coals - Polish standard classification.

wegiel metakokosowy - Polish term for a class of coking coal : See medium volatile coals - Polish standard classification.

wegiel ortokoksowy - Polish term for a broad class of coking coal : See medium volatile coals - Polish standard classification.

wegiel plomienny - Polish term for long-flamecoal : See high volatile coals - Polish standard classification.

W

wegiel sokoksowy - Polish term for a class of coking coal : See medium volatile coals - Polish standard classification.

weighted thickness (gewogene Mächtigkeit; puissance pondérée) - CEC Analysis of Terms: General arithmetic mean of the thicknesses of an area comprising zones of varying average thicknesses. The average thicknesses of the zones are weighted in proportion to their respective areas. (CEC Assessment of Coal Reserves, 1980).

wf - Germ abbreviation for wasserfrei : See dry - FRG standard definition.

Wichtestufe - German term: See density fraction - CEC Analysis of Terms.

wirtschaftlich bauwürdige Vorräte - German term: See economically recoverable reserves - CEC Analysis of Terms.

wmf - Abbreviation for wasser-und mineralstofffrei : See dry mineral-matter-free - FRG standard definition.

worked thickness (Abbaumächtigkeit; puissance totale) - CEC Analysis of Terms: Thickness of a deposit section which has been worked or is being worked (worked thickness of worked-out area) or which is expected to be worked (worked thickness of coal in place). It may exclude a section of the deposit or include sections of the surrounding rock. (CEC Assessment of Coal Reserves, 1980).

working loss (Verlustmenge Abbau; pertes á l'exploitation) - CEC Analysis of Terms: This is the quantitaive difference between layout reserves and working reserves (see sketch).

GEOLOGISCHER VORRAT COAL IN PLACE (GEOLOGICAL RESERVES)			
TECHNISCH WIRTSCHAFTLICHE VERLUSTE / TECHNO-ECONOMIC LOSSES	TECHNISCH GEWINNBARER VORRAT / TECHNICALLY RECOVERABLE RESERVES		
:::	VERLUSTMENGE PLANUNG / PLANNING LOSS	PLANVORRAT / PLANNING RESERVES	
:::	:::	VERLUSTMENGE ZUSCHNITT / LAYOUT LOSS	ZUSCHNITTSVORRAT / LAYOUT RESERVES
:::	WIRTSCHAFTLICHE VERLUSTE	VERLUSTMENGE ABBAU / WORKING LOSS	ABBAUVORRAT / WORKING RESERVES

(CEC Assessment of Coal Reserves, 1980).

working reserves (Abbauvorrat; réserves d'exploitation) - CEC Analysis of Terms: That portion of the technically recoverable reserves which can be made available to a consumer. Working reserves are therefore always expressed as a quantity measured at the surface, after preparation. Forward estimates may be made on the basis of experience if no more than an indication of tonnages is required. (CEC Assessment of Coal Reserves, 1980).

X

xyloid lignites (ligniti xiloidi) - Italian classification: - have a woody structure. They have a net calorific value from 1600 to 2200 kcal/kg on a run-of-mine basis within the general classification of lignites. (ENI-ENEL-FINSIDER Classification Systems for Coals, 1978).

Y

yield (Ausbeute; rendement) - CEC Analysis of Terms: An indication of the tonnage directly subjected to a winning process. This tonnage is calculated from the worked area times the average weighted thickness of coal times the average specific gravity of the coal. Quantities lost because of the winning process and losses in haulage and preparation are disregarded. (CEC Assessment of Coal Reserves, 1980).

Z

zekere - Flemish term: See **certain reserves** - Belgian definition.

Zugangsverhältnis - German term: See **access overburden ratio** - FRG standard procedure.

Zuschnittsvorrat - German term: See **layout reserves** - CEC Analysis of Terms.

Sources

ANSI/ASTM	-	American Society for Testing and Materials 1916 Race Street Philladelphia Pa. 19103 U.S.A.
AS	-	Standards Association of Australia Standards House 80 Arthur Street North Sydney New South Wales Australia.
Austrian Guidelines for Coal Deposit Assessment	-	Federal Ministry for Commerce, Trade and Industry Vienna Austria.
Belg. Coal Min. Ind. Exec.	-	Fedechar Rue de la Loi 99-101 BTE 7 1040 Bruxelles Belgium.
BS	-	British Standards Institution 101 Pentonville Road London N1 9ND United Kingdom.
Bull. geol. Surv. Taiwan	-	Geological Survey of Taiwan P.O. Box 968 Taipei Taiwan.
Canada: EMR Report EP	-	Energy Mines and Resources Department Government of Canada 580 Booth Street Ottawa, Ontario K1A 0E4 Canada.
Cdf Reserves Classification	-	Charbonnages de France Technical Services Directorate 9 Avenue Percier Paris 8 France.
CEC Assessment of Coal Reserves	-	Commission of the European Communities Rue de la Loi 200 B-1049 Bruxelles Belgium.

Dept. Mines, Rept. Coal Res. SA	-	Department of Mines The Government Printer Bosman Street, Private Bag X35, Pretoria South Africa.
DIN	-	German Institute of Standards Postfach 1107 - D1000 Berlin 30 F.R.G.
ENI-ENEL-FINSIDER Classification	-	ENI-ENEL-FINSIDER ENI Coleset Piazzale Enrico Mattei 1 Roma Italy.
Indian Standard Procedure for Coal Reserve Estimation	-	Indian Standards Institution Manak Bhavar 9 Bahadur Shah Zafar Marg New Delhi 110001 India.
ISO	-	International Organisation for Standardisation 1 Rue de Varembe Case Postale 56 CH-211 Geneve 20 Switzerland.
JIS	-	Japanese Standards Association Standards Department Agency of Industrial Science and Technology Ministry of International Trade and Industry 1-3-1 Kasumigaseki Chiyodaku Tokyo 100 Japan.
NCB	-	National Coal Board Hobart House Grosvenor Place London SW1 United kingdom.
NSW Code for Coal Reserves	-	Standing Committee on Coalfield Geology of New South Wales Geological Survey of New South Wales New South Wales Australia.

NVKS	-	N. V. Kempense Steenkolenmijnen Grote Baan 27 3530 Houthalen Belgium.
NZ geol. Surv. Bull	-	New Zealand Geological Survey Department of Scientific and Industrial Research P.O. Box 30-368 Lower Hutt New Zealand.
PN	-	Polish Committee for Standardisation Warsaw Poland.
Publ. Geol. Esp. Ingeominas	-	Instituto Nacional de Investigaciones Geologico-Mineras Carrera 30, No. 51-59 Bogota Colombia.
Queensland Coal Reserves Classification	-	Department of Mines Queensland Geological Survey of Queensland Mineral House 41 George Street Brisbane 4000 Queensland Australia.
Rhurkohlen-Handbuch	-	Rhurkohle A. G. Postfach 5 4300 Essen F.R.G.
UN CNR	-	United Nations Committee on Natural Resources United Nations NY 10017 New York U.S.A.
UN ECE	-	United Nations Economic Commission for Europe Palais de Nations Geneve Switzerland.
US geol. Surv. Bull.	-	United States Geological Survey Department of the Interior National Center, Mail stop 956 Reston Va., 22092 U.S.A.